KITCHEN & BATH
LIGHTING

KITCHEN & BATH
LIGHTING
Concepts, Design, Light

DANIEL BLITZER

TAMMY MACKAY, AKBD, NCIDQ, LEED GREEN ASSOCIATE

National Kitchen & Bath Association

National Kitchen & Bath Association
687 Willow Grove Street
Hackettstown, NJ 07840
Phone: 800-THE-NKBA (800-843-6522)
Fax: 908-852-1695
Website: NKBA.org

Published by John Wiley & Sons, Inc., Hoboken, New Jersey.
Published simultaneously in Canada.

For general information about our other products and services, please contact our Customer Care Department within the United States at (800) 762-2974, outside the United States at (317) 572-3993 or fax (317) 572-4002.

Wiley publishes in a variety of print and electronic formats and by print-on-demand. Some material included with standard print versions of this book may not be included in e-books or in print-on-demand. If this book refers to media such as a CD or DVD that is not included in the version you purchased, you may download this material at http://booksupport.wiley.com. For more information about Wiley products, visit www.wiley.com.

Library of Congress cataloging-in-publication data is available upon request.

ISBN 978-1-118-62772-3 (cloth); 978-1-119-12456-6 (ebk.); 978-1-119-12455-9 (ebk.)

Printed in the United States of America

10 9 8 7 6 5 4 3 2

Sponsors

The National Kitchen & Bath Association recognizes, with gratitude, the following companies whose generous contribution supported the development of *Kitchen & Bath Lighting*.

PLATINUM SPONSOR

WWW.HAFELE.COM

About the National Kitchen & Bath Association

The National Kitchen & Bath Association (NKBA) is the only nonprofit trade association dedicated exclusively to the kitchen and bath industry and is the leading source of information and education for professionals in the field. Fifty years after its inception, the NKBA has a membership of more than 60,000 and is the proud owner of the Kitchen & Bath Industry Show (KBIS).

The NKBA's mission is to enhance member success and excellence, promote professionalism and ethical business practices, and provide leadership and direction for the kitchen and bath industry worldwide.

The NKBA has pioneered innovative industry research, developed effective business management tools, and set groundbreaking design standards for safe, functional, and comfortable kitchens and baths.

Recognized as the kitchen and bath industry's leader in learning and professional development, the NKBA offers professionals of all levels of experience essential reference materials, conferences, virtual learning opportunities, marketing assistance, design competitions, consumer referrals, internships, and opportunities to serve in leadership positions.

The NKBA's internationally recognized certification program provides professionals the opportunity to demonstrate knowledge and excellence as Associate Kitchen & Bath Designer (AKBD), Certified Kitchen Designer (CKD), Certified Bath Designer (CBD), and Certified Master Kitchen & Bath Designer (CMKBD).

For students entering the industry, the NKBA offers Accredited and Supported Programs, which provide NKBA-approved curriculum at more than 60 learning institutions throughout the United States and Canada.

For consumers, the NKBA showcases award-winning designs and provides information on remodeling, green design, safety, and more at NKBA.org. The NKBA Pro Search tool helps consumers locate kitchen and bath professionals in their area.

The NKBA offers membership in 11 different industry segments: dealers, designers, manufacturers and suppliers, multi-branch retailers and home centers, decorative plumbing and hardware, manufacturer's representatives, builders and remodelers, installers, fabricators, cabinet shops, and distributors. For more information, visit NKBA.org.

Table of Contents

Preface

Welcome to *Kitchen & Bath Lighting*. As vision is our most significant sense, so lighting is critical to our perception of the world around us and to our effective performance, attractive appearance, and healthy emotions.

Lighting is one of the elements of design. Some might say it is *the* most important as without it we would not be able to see the other elements. It can enhance the look and feel of a space or detract from it if done incorrectly.

Kitchens and baths represent the residential spaces where lighting is used most often and most critically. The principles of lighting kitchen tasks and social areas often can be applied to other work and living spaces, while understanding how to light at a bathroom vanity can inform attractive lighting for conversations throughout the home.

ORGANIZATION

This book is intended both to provide a sound basis in the fundamentals of lighting and to guide in the application of lighting to the two most critical task spaces of the home. We approach kitchen and bath lighting in several broad categories familiar to designers.

In chapters 1 through 6 we focus on the fundamentals of lighting and discuss how we see materials, spaces, and each other; how to calibrate lighting for different tasks; and how to modulate lighting as we age.

Chapters 7 and 8 explore the importance of sustainable lighting and daylighting.

In Chapter 9 we cover schematic design by presenting a visual vocabulary for speaking about lighting and applying these ideas to conceptualizing lighting for kitchen and bath spaces.

Chapters 10 through 12 provide important information on choosing and comparing light sources and fixtures and in Chapter 13 lighting controls are discussed.

Chapter 14 explains the many aspects of design development including the selection of light sources, luminaires, and controls and the processes involved in locating equipment, calculating illumination, and addressing code compliance.

Chapter 15 covers the important topics of documenting the lighting design and communicating the design to the construction team and finally, in Chapter 16 we explain the process and critical issues involved with getting lighting built in the real world.

SOME CURRENT ISSUES

Three critical issues stand out today: changing lighting technology, sustainable design approaches, and lighting for older eyes.

If this book had been written 50 years ago, the lighting *principles*—how we see and how to arrange lighting—would have been largely the same. But, of course, technology in the twentieth century would have been significantly different. Indeed, if the book had been written even just a few years ago, light sources, luminaires, and controls would be considerably different.

Most notably, LEDs—light-emitting diodes—are rapidly changing what we light with, how light colors what we see, how it looks in our homes, and how it affects the natural environment. At the same time, control over lighting has become both more convenient and more sophisticated. The combination of digital light sources (LEDs) and digital controls promises a future of lighting that adapts readily to different needs, uses, and preferences. With the majority of the population working and playing on video displays of some kind, the way we light spaces has changed completely as well.

Around the globe, developing economies are trying to meet the fast-rising expectations of their populations. A peaceful world will need *sustainable lighting*—lighting that meets the human needs of today with the least impact on energy and other natural resources so as not to compromise future generations.

Young people often can work without any electric lighting. As people age, we typically use lighting more often and in greater quantity. This progression is inevitable, at least for most of us. And for much of the developing world, it is true not only for individuals but for the population as a whole. Providing for the range of needs required by residents of varying ages is a critical challenge for lighting design.

We pick up the strands of these issues throughout the book. Woven together, they help us to think about lighting holistically: who the lighting is for, what their needs and desires are, and how we can use design and technology to meet their needs and even exceed their expectations.

Dan Blitzer

I feel very grateful to have the opportunity to share my knowledge of and enthusiasm for the interior design industry, specifically kitchen and bath design. The National Kitchen & Bath Association's commitment to education in this area of expertise has been incredible. Johanna Baars, Publications Specialist, Lisa H-Millard, Course Developer, and Debby Mayberry, Learning and Development Implementation Specialist, have been instrumental in providing support and encouragement to me throughout this process.

The team at my firm and my family at home have also been supportive and excited about this new book added to the series. Thanks to all. I give my best to future kitchen and bath designers. I hope you enjoy your journey as much as I have enjoyed mine so far.

Tammy MacKay

Acknowledgments

The NKBA gratefully acknowledges the following peer reviewers of this book:

Kristen Arnold

Jeff Dross

Robert Dupuy

Cheryl A. Glazier, CKD, CBD

Corey Klassen, CKD, CBD

Anna Mahan

Natalia Pierce, AKBD

How We See

Light, how it enables us to see, and lighting terminology together provide the necessary foundation for understanding lighting. In this chapter, we begin to consider these fundamental concepts. In subsequent chapters, we investigate lighting fundamentals in more detail.

Learning Objective 1: Describe the physics of light and the physiology of the eye.

Learning Objective 2: Explain in plain language how we see.

Learning Objective 3: Recognize and use key lighting terms and metrics.

Learning Objective 4: Distinguish between perceived and measured illumination.

PHYSICS OF LIGHT

Light is the *energy* that enables us to see. Technically, light is part of the broad spectrum of electromagnetic energy and is defined as visually evaluated radiant energy (see Figure 1.1).

As you may recall from classes in physics, light exhibits the properties of both waves and particles. As a radiating wave, light can be described by its wavelength, which ranges from about 380 to 760 nanometers (billionths of a meter), the limits of human visual sensitivity. In the next chapter, we explain that describing light by its wavelength helps us to understand the interaction of light and materials. Later, when we look at light sources, we encounter the particle nature of light—especially in understanding LED technology.

A few observations:

- Light itself is invisible. We see it only when it interacts with materials (e.g., the filament of an electric light source, fabrics, or faces). More on this important idea shortly.
- Light can travel through some materials.
  - *Transparent* materials allow the passage of light without significant distortion so you can see the details of objects behind them (see Figure 1.2a).
  - *Translucent* materials allow light through but mix it up so that the details are obscured. (The entire object may be obscured, depending on the translucent material and the nature and location of the object.) (see Figure 1.2b).
  - *Opaque* materials block the passage of light altogether (see Figure 1.2c).

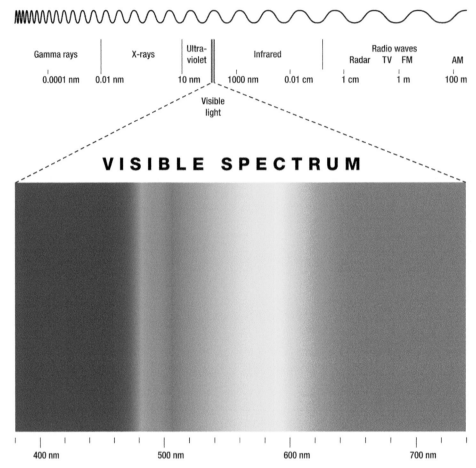

FIGURE 1.1 Light in the electromagnetic spectrum
Courtesy of Peter Hermes Furian

- Light changes direction when it reflects off surfaces or when it passes through materials, refracting (bending) or scattering (see Figure 1.3).
- Light that neither passes through nor reflects off materials is absorbed. Its energy becomes heat. Some light is absorbed in virtually every encounter with materials. Put your hand on the hood of a car that has been sitting in sunlight and *see* for yourself.

VISION

Although vision is not our oldest sense (we touch before we see), it dominates our perception. Basically, human vision is simple: Light interacts with objects; travels to, then enters, our eyes, where it is transformed into electrical signals; these signals travel neurological pathways to reach our brain, where they are interpreted into visual perception. Another way to express this basic process is by its four essential components (see Figure 1.4):

1. Light source
2. Object
3. Eye
4. Brain

We know a great deal about the physics of light and how it interacts with objects. We also know a great deal about the physiology of the human eye, how it receives light and creates neurological connections. We know considerably less about the complexities of how our neurological signals are combined with memory and interpretive algorithms into dynamic, three-dimensional perception.

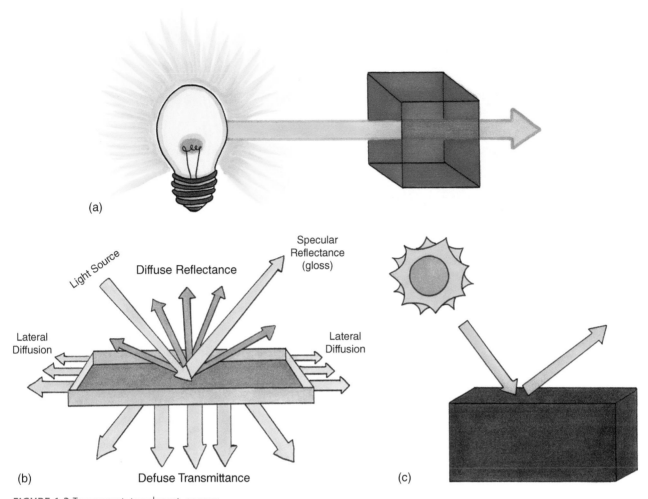

(a)

(b)

(c)

FIGURE 1.2 Transparent, translucent, opaque

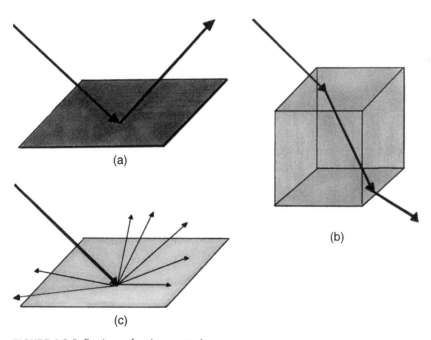

(a)

(b)

(c)

FIGURE 1.3 Reflection, refraction, scattering

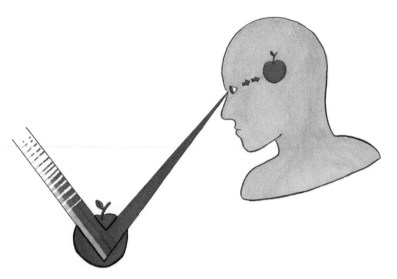

FIGURE 1.4 Light source, object, eye, brain

Pause for a moment to consider the following. The signals received on the two-dimensional "screen" of our retina are fundamentally ambiguous: Is the retinal image a small object close by or a large one at a distance? Yet, apart from some notable optical illusions, we see the world *un*ambiguously. This is only the most obvious example of our remarkable powers of visual perception. Good lighting can enhance these powers, while poorly designed lighting just makes seeing that much harder.

Contrast

Our visual system compares the incoming signals, searching for differences in light intensity and color. It does not measure or quantify them in technical photometric (light measurement) terms. Instead, the essence of how we see is the contrast between dark and light or among various colors.

Later in this chapter, we discuss how we measure light and all the technical terms associated with these quantities. When we do this, we also discuss the problems created by measurements that do not adequately represent perception.

Adaptation

Remarkably, our visual system operates effectively in a range of about 20,000:1, that is, from a bright sunny day to a starlit night. We manage to see in such a broad range by adjusting both the amount of light reaching the eye and the sensitivity of the photoreceptors. In darkened conditions, our pupils dilate to admit more light, and the eye's chemistry becomes more sensitive to the limited amount of light available. In bright conditions, in contrast, pupils contract, and sensitivity diminishes to avoid overload.

Adaptation takes time; it takes as much as 30 minutes to adapt to darkened conditions. Adapting to bright conditions takes less time. However, rushing the process (e.g., by emerging from a darkened theater to a bright afternoon) can prove painful.

Indoors, your vision adapts as you move from darker spaces to brighter ones and back again. Shifting your gaze from a brightly lighted task to a much darker surface also involves adaptation. Frequent and extreme adaptation can cause eye fatigue and discomfort.

Physiology of the Eye

The physiology of the eye helps us understand lighting—and how to design it for different applications and users of different ages and visual impairments. Take a moment to study the diagram of the human eye in Figure 1.5.

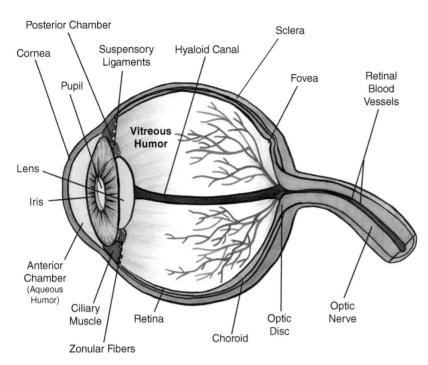

FIGURE 1.5 Diagram of the eye

We have already discussed an important function for the pupil: regulating the quantity of light received. The rays of light ultimately enter the eye through a lens that receives them onto the retina, which contains the photo sensors and connective neural networks that translate incident light into neurological signals. When the lens malfunctions (focuses improperly or simply loses clarity), vision is impaired.

Inside the eye, light travels through liquid from the lens to the retina. Impurities can disrupt light's passage and, with it, vision. Degradation in the retina (macular degeneration is one important example) also diminishes vision.

The retina contains three basic types of photoreceptors:

1. Those capable of detecting only the quantity of light, not its color, are called *rods* due to their shape. They are also capable of sensing very small quantities of light; we rely on them to see in the (near) dark. Located throughout the retina, rods also provide peripheral vision.
2. Those capable of detecting color are called *cones*. They require more stimulation than rods, so we enjoy very limited color vision in darkened conditions. Concentrated in an area called the *fovea*, cones provide the visual acuity to distinguish small tasks.
3. The third type of photoreceptor is not part of the visual system but detects light as part of our circadian, or 24-hour, clock system.

We return to the photoreceptors in more detail when we discuss color in Chapter 2, "Seeing Materials" (and to nonvisual photoreceptors when we discuss lighting for aging eyes).

Finally, notice how an overhanging brow protects the entire eye, limiting the glare from overhead sources of bright light—to some degree at least.

MEASURING LIGHT

Light emanates from a source and (some of it) arrives at an object. Light then leaves that object (reflects off or passes through it) and travels to the next object, and so on. Thus, if we want to measure light, we need to do so at the various points in its travel.

Lumen—the Flow of Light

Let's start with the source itself. A *lumen* is a unit of measure for quantifying the amount of light energy emitted by a light source. A typical light source in your dining room might emit 800 lumens, one in the laundry room might emit 2500 lumens, and one in the streetlight outside might emit 16,000 lumens.

LUMEN VALUES FOR VARIOUS LIGHT SOURCES

Light Source	Luminous Flux (lumens)	Typical Use
LED light bulb (12W)	800 lumens	Table fixture
Halogen flood (60W)	1100 lumens	Retail display
Linear fluorescent (28W)	2500 lumens	Office lighting
High-pressure sodium (150W)	16,000 lumens	Street lighting

Since light can flow in any direction, luminous flux (lumens) can be measured anywhere. The most common measurements are made at lamps (light sources) and fixtures (technically, *luminaires*, which are fixtures with light sources in them). You will find lumen ratings in lamp and luminaire catalogs, specification sheets, and websites, as well as on many lamp packages (see Figure 1.6).

FIGURE 1.6 Lamp package label
Courtesy of American Lighting Association

Candela—The Intensity of Light

If the flow of light (lumens) is concentrated into a tight beam, we say that it is intense (strong in that single direction). This is called *luminous intensity* and is measured in *candela*. Light sources and luminaires with distinct beams of light (whether concentrated as spots or diffused as floods) typically are measured in both lumens and candela.

With light sources that direct light into beams, candela measurements will vary significantly according to the angle at which the intensity is measured. With light sources that distribute light more or less evenly in all directions, the candela measurements will be very similar.

The word "intensity" has the common meaning of "strength." In lighting, luminous intensity has the technical meaning of "strength in a specific direction." We use the term "lumens" to measure the total flow of light, regardless of direction.

Footcandle/Lux—Light Falling on a Surface

When light falls on a surface, it is called *illuminance*. Technically, we measure the density of the illuminance that is the quantity of lumens falling on a surface. One lumen per square foot equals 1 *footcandle*. This is an imperial measurement. Its metric equivalent is *lux*. One lumen per square meter equals 1 lux. The conversion of footcandle (FC) to lux is 1 footcandle = 10.764 lux.

If 100 lumens fall on 1 square foot of countertop, we would measure that as 100 footcandles. If those lumens were spread over 10 square feet, we would measure it as 10 footcandles. The same comparison can be made for lux. If 100 lumens fall on 1 square meter of countertop, we would measure that at 100 lux.

Illuminance is measured with a luminance meter (also called an *illuminance meter*). The Illuminating Engineering Society has established recommended illuminance targets (in footcandles) for almost every room or setting.

Brightness—The Perception of Light

So far, none of our measurements represents what we *see*. Lumens and candela measure light at the source; footcandles (lux) measure light falling on the surface of an object. We see when light reflects from (or passes through) an object and reaches our eye. We can measure the amount of light detected by the eye from a surface at a particular angle; this is called *luminance*.

In practice, luminance measurements are cumbersome and costly, and so are rarely used in everyday design. Instead, we use the term *brightness* to express our perception of light, including the many non-quantified factors that influence our visual process.

MEASUREMENT VERSUS PERCEPTION

It is worth emphasizing that most of what we measure in lighting—the lumens flowing from a light source, the candela at the center of a beam of light, the footcandles falling on a surface—do *not* represent what we see when we look at a room, a task, or a person.

Our perception depends on the materials we are lighting, what we see around the space, how we are adapted to the brightness of the space, the color of the light, and other factors, which we discuss in later chapters.

What good are these measurements then? As you will learn, we apply the measurements of lighting to *predict* how well lighting will meet our objectives. By themselves, these measurements are of little use. But together with an understanding of how light interacts with materials, people, and task demands, they can help us judge how to provide appropriate lighting.

Using Lighting Terms

1. Look at a room in your home or place of business. Write a description of the lighting present using the terms "luminous flux," "intensity," and "illuminance."
2. Think about a recent day in your life. What is the brightest environment in which you can see easily? What is the dimmest?
3. What is the unit of measure for illuminance?
4. Look up the word "footcandle" online. What is its origin? Do the same for the word "lux."

SUMMARY

Light is the energy that enables us to see. It travels invisibly, and when it encounters a material, it reflects, refracts, or is absorbed by it. We see when light reaches the photoreceptors in our eyes, stimulating signals that are interpreted by our brain. Our visual system responds to contrast (rather than absolute levels of light) and adapts to variations in ambient brightness. We measure light in three widely used measurements: luminous flux (lumens), which refers to the flow of light; intensity (candela), which refers to the strength of light in a specific direction; and illuminance (footcandle/lux), which refers to the amount of light falling on a surface.

REVIEW QUESTIONS

1. What are the four elements in human vision? (See "Vision," page 2)
2. What is the role of contrast in vision? (See "Contrast," page 4)
3. What is the role of adaptation in vision? (See "Adaptation," page 4)
4. In what part of the eye are the photoreceptors located? (See "Physiology of the Eye," page 5)
5. What are the units of measure for luminous flux, intensity, and illuminance? (See "Footcandle/ Lux," page 7)

Seeing Materials

Light has distinctive qualities of color and form. Materials also present different qualities of color, form, and texture. In this chapter, we explore the different qualities of light and begin to see how those qualities interact with materials.

Learning Objective 1: Explain in plain language how we see color.

Learning Objective 2: Identify the primary colors of light, and explain why they are primary.

Learning Objective 3: Compare concentrated and diffuse light.

Learning Objective 4: Describe how light is reflected by matte and shiny surfaces.

Learning Objective 5: Distinguish between specularity and reflectivity.

COLOR

If light is invisible, how can it be said to have color? Observe the changing quality of skylight, the progression of sunlight from dawn to noon to dusk. Do you doubt that light indeed has color?

We explain this puzzle by dividing the subject of color into two parts:

1. The makeup of light itself—color in the energy that enables us to see
2. The interaction of light and material—color as we experience it

Color in Light

Look at the sky after a rainstorm. Pick up a prism and capture sunlight on its way to a nearby wall. You will see the familiar spectrum (see Figure 2.1) and perhaps remember ROYGBIV from school. ROYGBIV is the mnemonic for red, orange, yellow, green, blue, indigo, and violet, the colors we see as the sunlight separates by wavelength.

What we call *white light*—daylight and most electric light sources—combines light at different wavelengths. As light passes through refractive materials (e.g., droplets of water or a crystal prism), the constituent wavelengths are bent to differing degrees. Shorter wavelengths (blue) bend more than longer ones (red).

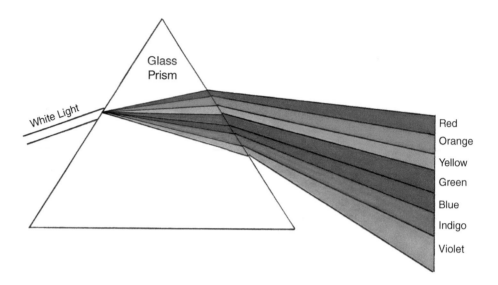

FIGURE 2.1 Spectrum from a prism

Refraction and Wavelength

You can see the varying refraction of wavelengths in the rainbow, where the inner bands are blue-violet and the outer band is red. In simple terms, this also explains why the sky is blue: Light entering Earth's atmosphere refracts slightly. Shorter blue wavelengths bend more than the others and so become visible as skylight.

Rainbows have been around since the dawn of time, but understanding what they mean with regard to light is relatively recent. You can replicate Isaac Newton's experiment by using two prisms. The first one separates light into different wavelengths; the second one reassembles those wavelengths to create white light and demonstrates that the "white" of light is nothing more than the combination of all colors of the spectrum.

Returning to our first question—What makes light appear in different tones of white?—it is the proportion of different wavelengths of energy. Afternoon skylight with its bluish tint, as seen in Figure 2.2, has a strong component of short wavelengths. At dawn or dusk, longer wavelengths predominate, and the light acquires a reddish tinge (see Figure 2.3). We call this the *spectral power distribution of the light source*.

FIGURE 2.2 Daylight in the afternoon
Photo by Dodge + Burn Photography

FIGURE 2.3 Daylight at dusk
Photo by Dodge + Burn Photography

Color Vision

In this chapter, we introduce the basic photoreceptors of the eye. Of these, it is the *cones* that play the key role in color vision. Cones respond to different wavelengths of light, enabling us to distinguish color. There are three types of cones:

1. Those most sensitive to long wavelengths of light (e.g., red)
2. Those most sensitive to medium wavelengths of light (e.g., green)
3. Those most sensitive to short wavelengths of light (e.g., blue)

Light arriving at the retina stimulates the cones according to the wavelengths in the light. Those signals combine in the brain to be interpreted as color.

Think about all of the colors that you can perceive or imagine all of the possible combinations of wavelengths in light. All of that input is received by just three photoreceptors. In other words, all perceived color is the result of three neural stimuli. We call these three colors of light—red, green, and blue—the *primary colors of light*. From them, we can create any other color. And we can create the sensation of white by combining them all (see Figure 2.4).

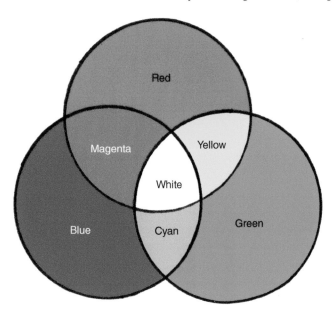

FIGURE 2.4 Blending the primary colors of light

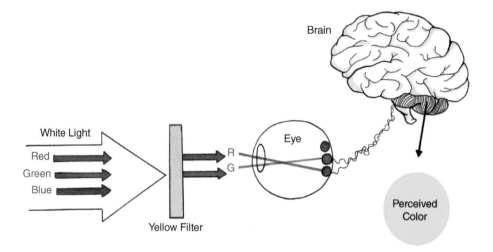

FIGURE 2.5 Diagram of color vision

White Light

Equal amounts of long, medium, and short wavelengths (red, green, and blue) produce a white light similar to afternoon daylight. Daylight at noon has more red and less blue; light dawn or dusk has even more red and less blue. We have more to say about the color and composition of daylight, sunlight, and skylight later in the book.

Color in Objects

You can see the fact that colors appear different in morning light than they do in the afternoon every day. The color we see in objects depends on the light under which they are illuminated. So, is there any intrinsic color in objects? Yes, and we call it the *pigmentation* to distinguish it from the color we perceive.

Pigments are simply selective reflectors of light. A red pigment reflects red light; a green pigment reflects green; a blue pigment reflects blue; and so on.

When you illuminate an object with white light, the object's pigments reflect selected wavelengths; those wavelengths enter your eyes and stimulate the cones, which send a mixture of red, green, and blue signals; finally, your brain interprets the signals as the color of the object (see Figure 2.5).

Here is the most important aspect of color vision: Your experience of color always combines the color (wavelengths) of light and the color (pigments) of the object. Considering light and object color independently makes little sense.

FORM AND TEXTURE

Light and shadow create the visual cues to understanding the form and texture of objects and materials. The principle is the same as the interaction of color in light and objects: The "form" of light interacts with the "topography" of objects to create our perception of their form and texture.

The form of light can be described as *concentrated* in a single direction or *diffused* across a wider area. Light that arrives at an object in a concentrated beam creates distinctive highlights and shadows according to the form and texture of the object (see Figure 2.6). Light that arrives in a diffuse wash minimizes those shadows and the perception of texture.

In front of a mirror, concentrated overhead light emphasizes facial features, such as eyebrows, nose, mouth, and chin, and creates the shadows we so dislike. Diffused illumination, in contrast, flattens form. At the mirror, this creates a softer and more pleasing aspect. Yet,

FIGURE 2.6 Concentrated light highlights the form and texture of this decorative wall.
Photo by Dodge + Burn Photography

over a bowl of fruit on the counter, that same concentrated light shows off form and reveals the intriguing texture of orange peel. Diffuse lighting might be called *dull* in this instance.

A single concentrated beam of light is called a *spot*. A series of concentrated beams aimed along a surface is called *grazing* light. Grazing light raises highlights and shadows along the surface and is especially effective at rendering surface textures. Think of rough stone or brick.

A single diffused beam of light is called a *flood*. When a series of diffused beams is aimed at a surface, it is called a *wash,* as in the term *wall washing.* Wash light minimizes highlights, shadows on the surface, and is preferred for lighting graphics (see Figure 2.7) and imperfect surfaces (think of taped and painted gypsum board).

FIGURE 2.7 Washing light design by Jaque Bethe Pure Design Environments, Bloomington, MN
Photo by Todd Buchanan

MATTE AND GLOSS

Matte and gloss are aspects of texture, but their textures are so fine that the impact of lighting differs considerably from that of more prominently figured materials.

Glossy surfaces—technically, known as specular or mirrorlike—reflect light at the same angle as it arrives. The angle of incidence equals the angle of reflection. Gloss finishes reflect light in a precise way, creating a clear reflected image in the surface when viewed from a particular direction. From other directions, no light is reflected and the surface appears dark.

Overhead lighting reflected in a bathroom mirror or lighting reflected in a polished countertop are common problems encountered with glossy materials.

Matte finishes, in contrast, reflect light diffusely, such that no image is visible. Walls finished with a flat paint and matte-finished laminate surfaces provide a soft and image-free appearance.

Specularity and Reflectivity

Specularity is not the same as *reflectivity*. Specularity indicates how bright objects will create images in the surface. Specularity can be considered as a range from fully diffuse (no image) to fully specular (perfect image), with most materials falling somewhere within the range. Reflectivity indicates how much light will bounce off the surface overall (see Table 2.1). Reflectivity is also a range, from zero (no light reflected; all light absorbed) to 100 percent (all light reflected; no light absorbed). The practical range for materials commonly used in homes is 5 to 95 percent.

TABLE 2.1 Specularity and Reflectivity

Material	Specularity	Reflectivity
Polished black marble	High	Low
Polished stainless steel	High	High
Dark red brick	Low	Low
White paint or laminate	Low	High

Looking at Light and Materials

Use your lighting journal to record your observations in this exercise. You can do this in your own kitchen or bath, in other rooms in your residence, or at work.

1. Select several objects representing a range of different colors.
   a. Observe them under different light conditions (daylight at different times of the day or different electric light sources).
   b. In your journal, identify the object and its dominant color, and describe how that color changes under different conditions of light.
2. Locate several surfaces with different textures.
   a. Observe them when lighted.
   b. In your journal, identify each surface, its dominant texture, and the form of the light (concentrated or diffuse) and describe the effect on the material.
3. Select an interesting three-dimensional object that you can move easily.
   a. Take the object somewhere so that it is lighted by a concentrated light source, then relocate it so that it is lighted by a diffuse light source.
   b. In your journal, identify the object and the two light sources, and describe the effect of the light on the object's form.

SUMMARY

In this chapter, we are beginning to understand the more technical aspects of how we view light and are better able to explain why we see the things we do. Lighting is an element of design that significantly affects the other elements, such as color, texture, and form. Already we can apply what we have learned to illuminate spaces more attractively and in ways that are more suitable for the function.

REVIEW QUESTIONS

1. Explain how we see in color. (See "Color Vision" page 11)
2. Describe the spectral composition of white light. (See "Color in Light" page 9)
3. What do we mean by concentrated and diffuse lighting? (See "Form and Texture" pages 12–13)
4. What is the difference between concentrated and diffuse light in their effects on objects and textures? (See "Form and Texture" pages 12–13)
5. How does light reflect from a matte surface? From a polished surface? (See "Matte and Gloss" page 14)

Seeing the Space and Each Other

In Chapter 2, "Seeing Materials," we looked at light, both its color and its intensity—concentrated or diffused. We focused our attention on materials and how our perception of them is influenced by these attributes of light. Now we broaden our vision to encompass entire spaces and the people in them. In particular, we consider how lighting affects the *character* of spaces, how we feel about—and in—them, and how we view the people we see in them.

Learning Objective 1: Explain in plain language color temperature and its significance.

Learning Objective 2: Understand how different patterns of brightness influence our experience of spaces.

Learning Objective 3: Discuss how light affects how we see people.

APPEARANCE OF LIGHT

Spectral power distribution (SPD), introduced in the last chapter, precisely describes the makeup of a light source in terms of the wavelengths of energy it comprises. However, SPD is a cumbersome way to apply light. Instead, we often describe light in two dimensions: by its appearance and by how well we think it renders colors. In this section, we focus on the appearance of the light.

Recall that light flows *invisibly* from its source to a surface. Once light reflects from—or transmits through—a material, our perception inextricably combines the light and the material. Therefore, to understand the appearance of light, we must look directly at the source (as unusual as that might be).

Seen in this way, one source of white light might appear golden, with a red-orange cast; we typically say that it appears warm. A different source might exhibit a slight bluish tint; we say it appears cool. *Warm* and *cool* are, of course, metaphors, perhaps drawn from nature, rather than measured attributes of the light.

Color Temperature

The terms "warm" and "cool" communicate a feeling well, but they are not precise enough to actually describe or distinguish a source of light. For this purpose, we would like to place

Color Temp.	Warm		Neutral	Cool	Natural	Daylight
Kelvin	2700K	3000K	3500K	4100K	5000K	6500K
Atmosphere	Soft, Comfortable, Relaxing		Efficient, Balanced	Clean, Efficient	Bright, Simulates Outdoors	Crisp, Refreshing, Energetic

FIGURE 3.1 Color temperature scale
Courtesy of American Lighting Association

light sources on a consistent scale that represents the appearance of light that is, we would like to assign a specific number to the appearance.

All solids, but particularly metals, glow as they are heated. At first, you cannot see the glow (and you also might burn your hand if you touch the solid)! But, as the object heats up, it begins to appear reddish. As it heats up still more, the glow becomes whiter, ultimately appearing blue-white. Thus, we can apply a scale of temperature to describe color. We call this, naturally enough, color temperature. That is, we describe the appearance of the light source by the amount of heat needed to create a glow of that color. The color temperature scale is measured in Kelvin (K) (see Figure 3.1).

Those red-orange tints of white light might have a color temperature of 2500 K. Ordinary incandescent light generally appears warm and has a rating of 2700 K. Fluorescent lighting in an office environment typically has a rating of 3500 to 5000 K; this lighting has more blue-green tints of white light and generally appears cooler.

Note the obvious pitfall: What we call warm-colored light has a *lower* color temperature than cool-colored light. Some designers prefer to associate warm and cool directly with Kelvin ratings and avoid the term "color temperature" when speaking with clients.

Color Temperature of Daylight

There is no single color temperature for daylight; it changes continuously. That is because the spectral composition of daylight changes over the course of the day (and the season and the location of the Earth in relation to the sun). The variation arises from the changing proportions of direct sunlight and indirect skylight as well as the atmospheric refraction of the different wavelengths.

Color Appearance and Color Rendering

The color temperature or Kelvin rating does *not* describe how a light source renders the color of materials to our eyes. The color we perceive results from the specific wavelengths in the light and the specific pigments in the material.

Nevertheless, because color appearance derives from the spectral makeup of the light source, we can draw some *general* conclusions about the effect of color temperature on our experience of colored materials. We return to the subject of color rendering when we discuss the selection of light sources in Chapter 10, "Choosing Electric Light Sources."

Color Appearance, Mood, and Culture

Generally, in North America, we associate warm-colored light with a relaxed mood and a leisurely, comfortable atmosphere (see Figure 3.2a). Cool-colored light, especially in high-brightness environments, suggests energy and activity (see Figure 3.2b).

FIGURE 3.2A Kitchen with warm atmosphere

Design by James Howard, CKD, CBD; codesigners Steve Levin and Sonja Willman, Glen Alspaugh Company, St. Louis, MO

Photo by Alise O'Brien Photography

FIGURE 3.2B Kitchen with cool atmosphere

Design by Sandra Tierney, CMKBD, CID; codesigner Doreen Owens, CKD, CBD, CID, Cabinets by Design, Escondido, CA

Observers have found association between the preferred color of light and climate, or geography. Areas with colder climates—Scandinavia, for example—seem to prefer warm-colored light (which also favors the wood that is widely used in regional home architecture). In warmer climates closer to the equator, people seem to prefer the cooling sensibility of higher Kelvin light. Cooler-colored light also appears to be the preference in Asia.

LIGHT AND OUR SENSE OF SPACE

Our eyes turn toward brightness. So it is natural that our perception of space should be strongly influenced by the distribution of brightness around the surfaces we see.

Interesting research performed in the 1970s examined how people respond to various arrangements of brightness in a space. Some of the conclusions suggest that:

- Lighting peripheral surfaces (walls, cabinetry) reinforces impressions of *spaciousness*.
- Nonuniform lighting reinforces impressions of *relaxation*. Peripheral, rather than overhead, illumination helps.
- Nonuniform lighting reinforces a sense of *intimacy*.
- Generally, nonuniform and peripheral lighting were preferred.

Note that this discussion emphasizes space as a volume rather than as a solid form or construction of materials. The location and character of the brightness influences our sense of the space more than the color or texture of the peripheral surfaces.

Peripheral Lighting

The term "peripheral lighting" refers to illumination falling on walls and other surfaces around the edge of a space. That is, with peripheral lighting, the periphery of the space feels brighter than the center. Uniform peripheral lighting might be described as wall washing. Nonuniform peripheral lighting can be created by a variety of techniques, such as a highlighted painting, an internally illuminated glass-front cabinet, or a wall sconce.

What distinguishes peripheral lighting most from basic overhead lighting is that it is mostly done with an indirect lighting effect (e.g., from a cove), as seen in Figure 3.3.

Nonuniform and Uniform Lighting

Nonuniform lighting creates contrasting areas of brightness and relative shadow. Uniform lighting, in contrast, creates a less-varied surface, although there can be noticeable contrast between two uniformly lighted surfaces, making the brighter one stand out.

General overhead lighting yields a uniform effect on the surfaces below (floor, work surfaces). Lights focused on a specific object or portion of a surface (e.g., an accent or local task light) create nonuniform effects.

Light and Spatial Impression

The most widely recognized researcher into the influence of lighting on spatial impressions is Dr. John Flynn, who worked at Kent State and The Pennsylvania State Universities in the 1970s with the support of the General Electric Lighting Institute. Flynn's experiments asked study participants to rate different lighting modes on various word-based scales. The results showed that there were consistent patterns of response to different lighting modes. Flynn looked at several contrasting modes:

- Uniform versus nonuniform
- Overhead versus peripheral
- Bright versus dim
- Visually warm versus visually cool

FIGURE 3.3 Lighting around the periphery
Design by Thomas R. Lamar II and Steve Garrison, Lamar Design Inc., Winter Park, FL

LIGHT AND PEOPLE

We consider how to *best* light people in Chapter 14, "Design Development." For now, we will consider what to avoid so that lighting will not produce an undesirable physical appearance of the people in the space.

More than Skin Deep

Skin pigments vary considerably depending on race and from person to person, but the blood underneath the skin is largely the same. Thus, most of us appear healthy when we are illuminated by light with a sufficient amount of long wavelengths (red) to show that our blood is red and oxygen rich. Light that lacks red content can make some people appear pale and generally unattractive.

Consider how we might detect changes in emotion by the blush, flush, or pallor of someone's demeanor. Or how a pale or yellowish countenance might suggest illness or at least discomfort. Since we often rely on these cues for social intercourse, it is important that lighting support, not hinder, these perceptions.

In other words, lighting should have enough red in the spectrum so people appear attractive and can see each other effectively.

Lighting Faces

People present wonderfully three-dimensional forms—especially our heads and faces, which protrude and recede most humanly (ears, noses, lips, chins, etc.).

The concentrated light that falls on us, particularly from directly overhead, can create dramatic highlights and shadows. Sometimes the result can be quite attractive; other times it can be somewhat grotesque. Generally, people look their best when lines and shadows are diminished, that is, when light is diffused rather than concentrated, arriving softly from several directions. Sharp shadows also diminish our ability to read expressions.

Looking at Spaces and People

Use your lighting journal to record your observations in this exercise. You can do this in your own kitchen or bath, other rooms in your residence, or at work.

1. Look at rooms that are lighted in different modes.
   a. In your journal, record your sense of spaciousness, relaxation, or intimacy.
   b. Now describe the primary lighting modes you observe. Do your perceptions agree with the text? (We all see the world differently!)
2. Observe people in different activities (e.g., working, dining, shopping), paying attention to their faces.
   a. In your journal, note the lighting in each space.
   b. Describe how it renders the faces of the people in the space.

SUMMARY

Although light flows invisibly, the appearance of a white light source can be described by its color temperature, measured in Kelvin. Sources with a high Kelvin rating appear a cool, bluish white, while those with a lower Kelvin rating appear a warm, reddish white. How spaces are lighted—particularly whether there is brightness on the periphery or overhead and whether the lighting is uniform or not—can influence impressions of spaciousness, relaxation, and intimacy. Finally, our perception of people is enhanced by generally soft quality of light with adequate red content.

REVIEW QUESTIONS

1. How does spectral power distribution differ from color temperature? (See "Appearance of Light" page 17)
2. What colors and materials tend to appear more vivid under warm-colored light? What colors and materials tend to appear more vivid under cool-colored light? (See "Color Temperature" pages 17–18)
3. What modes of lighting tend to increase the sense of spaciousness? Of intimacy? (See "Light and Our Sense of Space" page 20)
4. Why does concentrated light from overhead tend to be less flattering to most people? (See "Lighting Faces" pages 21–22)

Seeing the Work

Kitchens and baths are two of the most important work areas in the home. This chapter discusses how we see different tasks, how we determine how much light is needed to see them, and how to recognize challenging impediments to effective workspace lighting.

Learning Objective 1: Evaluate the key factors for task visibility.

Learning Objective 2: Recognize horizontal and vertical tasks.

Learning Objective 3: Understand how to choose illuminance levels.

Learning Objective 4: Identify the key impediments to task visibility.

Learning Objective 5: Discuss the issues of seeing tasks together with seeing materials, spaces, and people.

TASK VISIBILITY

Think of a basic kitchen task, such as reading a recipe from a cookbook while cooking. Open the book to the recipe and place it on the counter (see Figure 4.1).

- The pages are white paper; the print is black ink. The *contrast* is quite high, which makes the text stand out.
- The text is set in 12-point type of a comfortable font design, and you are viewing from about 20 inches away (distance from eye to page). Together, these factors determine the "visual size" of the task, which appears large enough that you do not move closer.
- You are in the midst of preparing a dish on the nearby range and forgot exactly in what order to add some ingredients to the fast-cooking food. So you are pressed for time and need to speed up your reading, which seems harder than you would have thought. You turn on the light.
- Your mother walks over to help out. Finding it difficult to see the task at her age, she turns on additional light.

You have just considered the key factors affecting task visibility:

- Task contrast
- Visual size
- Time, speed, and accuracy
- Age

FIGURE 4.1 Reading cookbooks is one of the many kitchen tasks for which proper lighting levels must be planned.

Design by Peter Ross Salerno, CMKBD, codesigners Marsha Thornhill and Kimberly Hill, CKD, CBD, Peter Salerno, Inc., Wyckoff, NJ Photo by Peter Rymwid Architectural Photography

Task Contrast

Task contrast is the difference between the reflectance of the task and that of its background. The bigger the difference, the higher the contrast and the better the visibility.

Printed text on a page or label is an obvious example, generally, of high contrast. Here are some others:

- Dirt on a countertop, bathroom fixture, or the floor
- Food stuck on a utensil or pot
- Stubble on your cheek
- Food sautéing in a pan
- Solids dissolving in liquids
- Blemishes on skin

FIGURE 4.2 The shattered glass contrasted against the ground is an example of task contrast.
Courtesy of Creative Commons-Share Alike 3.0

- Images in a photograph
- Text from a printer
- Broken glass on the floor or ground (see Figure 4.2)
- Broken china on the floor or counter

Contrast also characterizes materials with different colors, what is called *color contrast*. A green bowl will stand out in a dark wooden cabinet (about the same reflectance), where a wood bowl will be harder to see.

Visual Size

As noted in the cookbook example, visual size combines the actual size of the task and the distance from which it is viewed. Bend over the cookbook, and the text appears larger. Notice how close you hold a box of food to read the ingredients.

If you want to pick out a mug from a cabinet shelf, the relevant size is the form of the mug (as opposed to a glass, for example) (see Figure 4.3). However, if you want to pick out a mug with a particular saying printed on it, the relevant size is that of the print. (Note that mugs of different colors are very quickly distinguished.)

Here are some examples of visual size and different tasks:

- Picking up a pot from a drawer
- Reading a large-print cookbook
- Reading ingredients on a food label
- Reading directions on a medicine label
- Stubble on your cheek
- An eyelash
- A splinter in a child's finger
- The hole in a hook-and-eye fastener
- A toy on the floor
- A sliver of broken china on the floor
- A particle of food at the bottom of a pot

FIGURE 4.3 Example of task size
Courtesy of Wellborn Cabinet Inc.

Time, Speed, and Accuracy

It takes about 1/10 of a second to see—that is, for the light reaching the retina to create an electrical signal, for the signal to pass through our neural networks, and for those networks to resolve the signals into an unambiguous image.

The faster we try to resolve the visual signals being sent from our eyes to our brain, the more we guess at what we see. "Guess" here means associate the brain stimuli with previous experiences stored in our memories.

The quicker we guess, the more likely we are to make a mistake. Thus, speed and accuracy are related. The longer we look at a patterned floor or countertop, the more likely it is that we will discern the pieces of food that fell on it (see Figure 4.4).

FIGURE 4.4 Example of task visibility

Time and task visibility obviously play critical roles in determining lighting for transportation and roadways, industrial machinery, and sports facilities. What about speed and accuracy at home?

- Reading directions on medicine while a child is hurt
- Reading directions while food is cooking
- Judging the condition of food while it is cooking
- Judging cleanliness while wiping

Age

We discuss how aging affects vision in detail in Chapter 5, "Seeing as We Age." For now, we simply say that older eyes receive significantly less light at the retina than do younger eyes. All other factors constant, people of 55, on average, experience *half the light* as people of 25.

The recommended levels of illumination that we describe later in this chapter in Table 4.1 are intended to serve people from 25 to 65. Where the primary users of the space are over 65, the recommended levels *double*. And lighting needs may be even greater for people age 80 and beyond.

Luminance

Simply put, luminance is the brightness of the task seen from our point of view. Understanding luminance as a concept helps us to understand and apply light more effectively, even if we do not commonly use the term in everyday work.

Importantly, luminance is not the same as illuminance, which is the light arriving at the task, and is measured in footcandles or lux. Although you can measure and calculate luminance, we rarely do so in residential applications.

Technically, luminance measures the strength of light coming from an object, as seen from our point of view. Luminance represents light reflecting off or transmitted through an object in the direction of our eye. Thus, luminance involves the *source* of light, the *task*, and the *eye*, and how they relate to each other. The luminance of a light source that is intrinsically luminous does not depend on the task.

Task reflectance and specularity (finish) affect how much of the incident light reflects toward the eye. Materials with a matte finish reflect light diffusely so a portion of the incident light reaches the eye, creating task luminance. Specular materials reflect light according to the angle at which the light arrives. Depending on the angle, task luminance will be very high or very low. In either case, this could be problematical.

The term *illuminance* refers to the light striking an object (technically, the density of light on the surface, or lumens per square foot or square meter, depending on whether you are using footcandles or lux). Neither the reflectance nor the specularity of the object has any effect on the light striking it.

Task visibility is a function of all these factors (task contrast, visual size, time, speed, and accuracy, age, luminance), which are independent of each other. That is, a change in task *size* does not itself change the contrast of the task, the age of the viewer, or the luminance. When one factor diminishes task visibility (e.g., low contrast), we try to compensate by improving another factor (e.g., enlarging the task, slowing down to view it longer, or applying more light).

VISUAL PERFORMANCE

We light a task—that is, we increase task visibility—to enhance the ability to perform the task. In the cookbook example, considering just the contrast and size of the task, there was sufficient task visibility *for you* without turning on the light. When reading time was reduced (and task visibility with it), you compensated by adding light. And, when the task was performed by older eyes, still more light was needed.

In both cases, adding illumination raised task visibility to levels where it was possible to read more easily. If task visibility is low, adding illumination brings a rapid improvement in performance. Improvement diminishes, however, once the task becomes sufficiently visible. After that point, increasing task brightness has minimal additional impact. Thus, it makes sense to ensure that tasks reach a plateau of visibility for the people performing them. Illumination recommendations are based on this principle.

ILLUMINANCE

The Illuminating Engineering Society of North America (IESNA) publishes *The Lighting Handbook*. The handbook provides a technical framework, a design guide, and approaches to various applications, including residences. It also provides recommended illuminance (footcandle or lux levels) for the different tasks within the application.

Applying the concepts of task visibility and visual performance, the IESNA has created 25 application categories, organized by contrast, size, time, and accuracy. The categories (denoted in the handbook as A to Y) range from the least demanding (orientation in a darkened environment), to the most demanding (some health-care procedures).

The task categories require different luminance for optimal visibility and performance. This luminance value (light reflecting from the task), in turn, requires a certain amount of light falling on the task—the illuminance.

Footcandles

The standard U.S. unit of measure for illuminance is the footcandle, equal to 1 lumen flowing over 1 square foot of surface. Twenty footcandles is the flow of 20 lumens over 1 square foot. The metric unit of measure is lux, equal to 1 lumen flowing over 1 square meter.

Footcandles and Lux

One meter converts to 3.28 feet, about 39 inches. One square meter equals 3.28 × 3.28, or 10.76 square feet. So, 1 footcandle equals 10.76 lux (do the math). For products sold in North America, often both lux and footcandles are noted.

Illuminance Levels

The recommendations for illuminance in kitchens and baths draws on the ideas and practices in *The Lighting Handbook*. We look more closely at illuminance levels in the context of the overall lighting design in Chapter 14, "Design Development." For now, simply consider the range of illumination for different tasks and what that suggests for lighting design (see Table 4.1).

TABLE 4.1 Task Categories

Illuminance FC (lux) Age 25–65	Typical Kitchen Tasks	Typical Bathroom Tasks
3 (30)	Living room (for reference)	
5 (50)	General	Frequent storage
10 (100)		Toilet
20 (200)	Breakfast area	Vanity
30 (300)	Cooktop and sink	Grooming
30 (300)	Reading normal print	
50 (300)	Preparation counters	
50 (500)	Reading 6-point print	

Measuring Illuminance

You can measure illuminance with simple footcandle meters, which can be purchased online or from lighting suppliers.

For purposes of this book, even an inexpensive meter will help you correlate your observations to measured values. Professionally, a more reliable meter (accurate to within 10 percent and corrected for human visual sensitivity) makes more sense.

A few tips for using the meter:

- Check the settings to see that you measure consistently.
- Place the meter's sensor flat on horizontal or vertical surfaces. Tilting the sensor distorts the measurement.
- Place the readout where you can see it, and step away so that your body neither shadows nor reflects light.

Horizontal and Vertical Tasks

Tasks performed on tables, countertops, and island surfaces are seen in the horizontal plane. These are called *horizontal tasks*. Typically, light reaches the tasks from above, spreads over the surface, and reflects easily to your eye (see Figure 4.5).

Tasks performed at a mirror and in cabinets, in contrast, are seen in the vertical plane (see Figure 4.6). Light from *above* spreads over a larger surface and reflects differently from on horizontal tasks. Typically, less light from above reaches your eyes. Light from in front (e.g., alongside a mirror) illuminates vertical tasks more effectively.

FIGURE 4.5 Example of horizontal task
Courtesy of Blum, Inc.

CHALLENGES TO TASK VISIBILITY

Challenges to task visibility and visual performance typically arise from three sources:

1. Tasks
2. Viewer
3. Lighting quality

FIGURE 4.6 Example of vertical task
Courtesy of Blum, Inc.

Small Size and Low Contrast

Tasks of small size and low contrast present common challenges to lighting in kitchens and baths. These include badly printed reading material, directions crammed into the small space available on medication or food containers, hard-to-reach areas to be cleaned (and therefore areas that are hard to see up close), visual cooking (as opposed to time- or temperature-controlled cooking), and many others. In most cases, better illumination addresses the problem.

Viewer

By "viewer," we mean the person performing the task, whose eyes may need better illumination, even for tasks of only modest difficulty. We say "better" rather than "more" because increasing the amount of light alone may create glare. For older eyes, glare may be as troubling as inadequate illumination. We look into this challenge more in Chapter 5, "Seeing as We Age."

Lighting Quality

By itself, the recommended illuminance level does not ensure effective task visibility and performance. *Quality* is as important as quantity; indeed, light quality usually is *more* important.

Although quality can be a slippery topic, three aspects of quality as it relates to task visibility merit discussion at this point:

1. Direction
2. Uniformity
3. Color

The *direction of the light*—where it originates relative to the task—is a major problem, particularly at kitchen counters. Light that originates in front of the task and reflects toward your eye can create reflected glare and veiling reflections. Reflected glare is excessive, distracting, and uncomfortable brightness reflected to your eye. The term "veiling reflections" refers to light that reduces task contrast.

Both reflected glare and veiling reflections are due to the location of the light source and the direction from which the light flows (see Figures 4.7 and 4.8). Both problems are exacerbated by specular finishes on the task or in the task area. Correctly locating lighting equipment in task areas helps to resolve these challenges.

FIGURE 4.7 Reflected glare

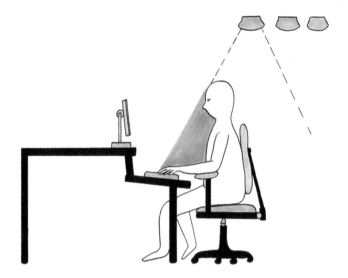

FIGURE 4.8 Veiling reflections

The *uniformity of the light* over the task area ensures that wherever you work in that area, you will have adequate illumination (see Figure 4.9). On a task surface, uniformity typically is measured by the ratio of maximum to minimum illuminance. Such a ratio should not exceed 3:1; closer to 1:1 is better.

Close spacing of light sources and a diffuse rather than a concentrated quality of light, generally produce effective results.

The *color of the light* source affects task visibility when the spectrum of the light does not match the spectral reflectance of the task materials. Imagine applying cosmetics under a light source that dulls the appearance of powder, nail enamel, or mascara. A similar challenge can

FIGURE 4.9 Uniformity of light for the island is provided by three large pendants centered over it.
Design by Carl Bruen, CGR, codesigner Robin Bruen and Debbie Kerr, CKD, Bruin Design Build, Morristown, NJ Photo by Wing Wang

arise when trying to determine whether a dish is fully cooked. Avoiding light sources with a poor color spectrum is the key to addressing this problem.

Looking at Task Illumination

Use your lighting journal to record your observations in this exercise. You can do this in your own kitchen or bath, in other rooms in your residence, or at work.

1. Look at four to six work surfaces. Choose one that seems most typical in terms of brightness and number it as (1). Number the others from (2) on.
2. List the spaces in your journal, leaving room for two columns to record illuminance levels. Column A records your observation; column B records your measurement.
3. Arbitrarily assign an illuminance value of 100 to surface (1).
4. Compare the other surfaces to (1), and assign an illuminance value to them, based on how bright they appear compared to (1). For example, a surface that appears half as bright would rate 50; one that appears one third brighter would rate 133. Write these values in column A.
5. Use a light meter to measure the illumination at each surface.
6. How do your observations compare to your measurements?

SUMMARY

Task visibility involves five factors:

1. Contrast of the task
2. Size of the task
3. Time available to see it
4. Age of the viewer
5. Luminance of the task

Lighting improves task visibility, but the improvement diminishes sharply once the task is easily seen. *The Lighting Handbook* of the IESNA categorizes visual tasks according to difficulty and recommends illuminance levels for various tasks. Effective task lighting requires good quality in terms of the direction, uniformity, and color of light as well as appropriate illumination.

REVIEW QUESTIONS

1. Find two examples of tasks with high and low contrast. (See "Task Contrast" pages 24–25)
2. Find two examples of tasks of large and small size. (See "Small Size and Low Contrast" page 31)
3. Name two tasks where speed and accuracy are very important and two where they are not so important. (See "Time, Speed, and Accuracy" pages 26–27)
4. Discuss three differences between luminance and illuminance. (See "Luminance" page 27, and "Illuminance" page 28)
5. Name three common challenges to task visibility and performance. (See "Challenges to Task Visibility" page 30)

Seeing as We Age

Vince Butler, CGR, GMB, CAPS

Vision, like our other senses, is affected by age. The changes that occur in the eyes with normal aging, as well as age-related eye disease, create several challenges for the lighting designer. A person 60 years of age receives only about 40 percent of the light at the retina as a 20-year-old receives. The simple solution might seem to be increasing the amount of light to compensate. Unfortunately, this solution often has negative consequences, such as creating glare and harsh contrasts. A better understanding of the effects of age on vision will assist in recommending and delivering solutions that can provide a safer, more comfortable, and more effective living environment.

> *Learning Objective 1: Understand the ways vision changes as we age.*

> *Learning Objective 2: Recognize age-related eye disease and how it affects vision.*

> *Learning Objective 3: Recommend lighting techniques that compensate for age-related vision changes.*

> *Learning Objective 4: Use creative lighting design to improve safety for older clients..*

HOW VISION CHANGES WITH AGE

As we discussed in Chapter 1, "How We See," the lens at the front of the eye is responsible for focusing light on the retina. As we age, the lens hardens and yellows. The tiny muscles that are responsible for moving the lens to focus are not able to overcome this hardening. The result is presbyopia, sometimes referred to as aging eye condition. The eye can no longer focus on close objects, and most people find it difficult to read small print or do close, delicate tasks without the aid of magnifying lenses or reading glasses. These changes also cause light to scatter in the eye, making the eye less effective at adjusting to glare and changing light conditions. It may be harder to distinguish colors, as they appear muted. Yellowing of the lens can make discriminating colors in the blue, purple, and brown color ranges especially difficult.

In addition to these normal changes, several age-related eye diseases can affect vision and need to be considered by the design professional. Low vision is defined as the loss of eyesight that makes everyday tasks difficult. It is most common for those over 65 years of age and may be the result of disease, injury, or other defects. Its effect on clients' ability to safely and effectively function within the home can be significant. Here are several common age-related eye diseases:

- *Cataract* is a blurring of the lens usually caused by protein clumping. It results in blurred vision, enhanced sensitivity to glare, poor color rendition, and decreased night vision. It is so common that by age 80, half of all people either will have a cataract or have had surgery to repair one. Because cataracts are common and usually correctable, now they are often considered a normal aging change rather than a disease.
- The term "glaucoma" refers to a group of diseases that affect the optic nerve and damage side (or peripheral) vision. Glaucoma is caused by high fluid pressure within the eye. If diagnosed early, it can be treated effectively, and permanent vision loss can be prevented. Otherwise, it can result in blindness. It is sometimes called the silent thief of sight because the symptoms develop gradually and often go undetected until the condition is quite severe and treatment is ineffective. Early detection through regular screening is critical. Glaucoma is most common in those over 60 years of age, in African Americans, in Hispanics over 40 years of age, and in those with a family history.
- *Diabetic retinopathy* is one of the leading causes of blindness in adults. It is a complication due to diabetes, which damages the blood vessels in the retina. It can also increase a person's risk of cataracts and glaucoma. According to the National Eye Institute, between 40 to 45 percent of people with diabetes have some stage of diabetic retinopathy. As with other eye diseases, early detection and treatment is critical to preserving sight.
- *Age-related macular degeneration* is another leading cause of blindness in those 50 years and older. It causes gradual loss of sharp, central vision rather than complete blindness and is caused by damage to the macula, a small spot at the center of the retina. This makes doing most routine tasks more difficult or impossible. Risk factors include age, race (more common in Caucasians than blacks or Hispanics), smoking, and family history.

Lighting Solutions for Aging Eyes

There are four basic areas to address in considering lighting for aging vision, in regard to both normal changes and eye disease:

1. Increased intensity
2. Uniform, balanced lighting
3. Control of glare
4. Help with color

Let us look at each of these and how to address the client's needs as well as how to best accomplish them.

Increased Intensity

Increased intensity is often straightforward, such as increasing lumens at the fixture and adding additional lighting throughout the design space. However, it is important to consider its effect on the other goals, such as the need for uniformed, balanced lighting and minimizing glare. If we just increase the intensity without regard to these concerns, we are likely to create a very bright, but worse overall, condition for clients. Place fixtures and light sources to create the desired intensity while not creating "islands" of brightness mixed with dark shadows. Look for a mix of ambient, task, and accent lighting that works in harmony with clients' use of the space. The best solutions usually include good lighting control with dimmers to allow for changing the intensity to suit the activity and time of day.

Uniform, Balanced Lighting

Since older eyes do not adjust to changing light conditions as quickly, it is important to consider the uniformity of lighting both within a space and between adjacent spaces. Try to balance the lighting throughout a room and compensate for extra-bright areas while

FIGURE 5.1 Floor-to-ceiling windows and glass doors provide bright natural light during the day. Lights in the exhaust hood provide additional lighting for the island.

Design by Claire Reimann, AKBD, Jason Good Custom Cabinets, Victoria, BC Photo by Joshua Lawrence

eliminating dark corners. If there is a bright natural light source, such as a skylight or large window, include additional lighting to provide balance during the day and to replace the natural light at night (see Figure 5.1).

This is a great example of why it is important to visit the site at different times of the day to observe how natural light changes and impacts the design space.

The lighting plan should try to eliminate any situation where there is a sharp difference in brightness while moving from one area to another within the home. This is not only uncomfortable but can increase the risk of falling. Motion-activated switches located just outside a room or corridor can anticipate an occupant's entry and illuminate the space in advance. Motion-activated switches are especially useful near stairs, both inside and outside the home. People often have their hands full or may be using a cane or walker when negotiating stairs and may not always turn the light on if the switch is inconvenient or unreachable. On exterior stairs, ensure that security lighting does not create a dangerous glare while negotiating the stairs. Position high-intensity lights away from stairs and landings, and illuminate those areas with softer, ambient light that outlines transitions clearly. Also, consider low-level lighting that lights the stair surface while not directing the light upward.

Older eyes fatigue more easily, especially when looking at areas of sharply contrasting brightness. For instance, driving at night is often uncomfortable with bright headlights appearing from the surrounding darkness. In the home, similar conditions can be avoided by placing accent lighting to balance a bright computer or TV screen. Clients will be more comfortable reading with a good task light in a softly lit room as opposed to having the surrounding room dark. Good lighting design with conveniently accessible controls will encourage use and minimize older clients' distress from eye fatigue.

Control of Glare

In order to eliminate glare, it is important to select fixtures or locate light sources so the bulbs are not exposed to view. Ambient lighting directed at the ceiling and reflected into the room can be quite effective. Lights over wall cabinets or under toe-kicks in kitchens and bathrooms will provide illumination without creating glare. This is referred to as indirect lighting. Task lighting that is shaded from view, such as under-cabinet lighting with shrouds or cabinet valances, should be used to illuminate all work areas (see Figure 5.2).

FIGURE 5.2 Under-cabinet task lights illuminate the counters while pendants add task lighting for the island.
Design by Anastasia Rentzos, CKD, CBD Andros Kitchen & Bath Designs, Mississauga, ON Photo by Averill Lehan/PAI

Consider the locations where clients will be reading or engaging in a hobby or craft. Remember to address the need for good task lighting at the sink in both the kitchen and the bathroom. In addition to the obvious tasks at this location, older clients often take prescription medication here. They will appreciate the additional light while reading labels and directions.

Glare can be the result of light reflected off a shiny surface. In addition to your lighting decisions, you will have to consider the choice of paint sheen and material surfaces. For older clients, highly polished stone and wood surfaces may not be the best recommendations. If these cannot be avoided or changed, then the lighting must be selected and located to minimize this reflected glare. Window treatments can be effective at diffusing natural light and can be used in conjunction with ambient lighting to balance the necessary intensity while reducing or eliminating glare. Lighter colors and lower-sheen paint, especially on ceilings, will help to reflect light without creating glare.

Help with Color

Color choice assistance is provided by selecting and locating lighting where it will be of particular assistance in clients' daily routines. Closets, dressing areas, laundry and craft rooms should have good illumination with task lighting that provides the required brightness at the point of use. In addition, the light source should have a color rendering index of at least 80 (preferably 90) or above.

Yellowing of the eye lens is common as we age, and its effect on color is exactly what you would expect if you looked through a yellow filter. It tends to scatter and absorb blue light. This makes it hard to distinguish between shades of blue, green, and violet. It can also make it hard to see soft contrasts between objects and their background, such as countertops and stairs. Using warmer colors, such as yellow, red, and orange, to highlight transitions will improve use and safety.

LIGHTING AND SAFETY

Lighting plays a critical role in preventing accidents. Falls are a leading cause of injury and death for older adults, and about one third are the result of environmental hazards. They often can be prevented with improved lighting that assists clients in seeing better and in maintaining good orientation and balance. Dimmable ambient lighting and automatic nightlights create a safe path and eliminate disorienting transitions between light to dark areas. Proper lighting encourages better nutrition by making it easier to prepare meals and keep the kitchen clean. Task lighting makes it easier to read labels and identify products, which helps prevent poisoning or prescription drug mishaps.

Night-Lights

All clients will benefit from attention to night-lighting within the home. Older clients especially will appreciate the convenience and safety provided by well-planned, automatic night-lighting. In addition to decreased vision, older clients suffer a higher incidence of insomnia, more frequent trips to the bathroom during the night, and mobility impairments that put them at higher risk of falling. Consider clients' floor plan critically. How do they move from the bedroom to bathroom and living areas? Are there stairs or other difficult transitions that need to be navigated? Automatic lighting that provides low-level, glare-free illumination of those pathways should be specified. Night-lights that are amber in color cause less disturbance of the circadian rhythm. Use either photocell or motion-activated controls so clients do not have to turn these lights on and off. Low-voltage rope lighting can be very effective under toekicks, handrails, and stair nosings (see Figure 5.3).

Some older clients may experience disorientation and confusion when first waking. Providing soft illumination of the entire room will assist with recognition and orientation. It also can help with balance if linear shapes, such as door frames and corners, are clearly visible to provide visual references.

FIGURE 5.3 Lighting under the toe-kick is effective in this kitchen designed with dark cabinetry and hardwood floors.
Design by Tracey Scalzo, CMKBD, Eurotech Cabinetry Inc., Sarasota, FL Photo by Tom Harper Photography

Special Considerations for Safety and Convenience

Lighting design offers many opportunities to improve client safety and convenience. The good news is that usually these benefits can be accomplished without compromising aesthetics or adding significantly to the overall budget. Addressing safety at every opportunity may not be a high priority for clients, but it should be considered a professional responsibility. Convenience improvements increase the use and enjoyment of the home and often enhance safety as well. Here are a few suggestions:

- *Use multiple fixtures at all hazardous locations.* These include stairs, bathrooms, and entrances. They provide redundancy in the event of a lamp burnout.
- *Illuminate corridor walls*, as older people often walk near walls rather than in the middle of corridors as this makes them feel safer.
- *Consider how lamps will be replaced.* Try to avoid the need for ladders, step stools, and the like. Where fixtures are harder to reach, select lights sources with long life expectancy.
- If possible, *select fixtures that use similar lamps to simplify replacement stocks*.
- *Utilize automatic lighting controls.* They are convenient and energy efficient ways to turn lights off when not in use. A good example is closet lights controlled by motion-activated switches.

- *Use lighted switches.* They assist with locating controls when entering a dark space and provide remote indication of lights being on or off.
- *Utilize wireless controls.* Such controls can assist mobility-challenged clients and reduce the number of transitions necessary throughout the day. This reduces fall risk, conserves human energy, and enhances convenience.
- *Avoid flickering or flashing lights* that can create disorientation and loss of balance. Upgrade existing fluorescents to electronic ballasts to eliminate the flickering. Watch for movement from ceiling fans that can interfere with lighting sources.
- *Provide well-lit, high-contrast house numbers* that are easily visible from the street to assist first responders in locating the home. It will also assist delivery people and visitors in finding the home.

Partnering with Health Care Professionals

When working with clients with special vision challenges, it is especially helpful to collaborate with a health care professional who specializes in low-vision assistance and therapy. Occupational therapists (OTs) are often best suited to assist in identifying solutions for home modifications to assist these clients. They work with clients to develop techniques and use special aid devices to overcome their limitations and accomplish everyday tasks. Ask clients if they have worked with a home health care provider or OT in the past and, if so, whether you can contact the person. If not, check your state licensing agency or OT programs at local universities for referrals. By working together, you will have access to in-depth information about clients' current condition, future prognosis and expectations, as well as additional design ideas to better serve their special needs.

Experiencing Low Vision

1. Smear a dime-size amount of petroleum jelly in the center of each lens of a pair of inexpensive sunglasses. This will simulate the effects of macular degeneration. Wear these around the house and try to do a few common tasks, such as reading, preparing a snack, or watching TV.
2. Now smear the entire lens with petroleum jelly to simulate cataracts. Try a few different tasks, and explore what type of changes to the lighting environment would assist you.
3. Identify locations within your home or business where a vision-impaired person would be at increased risk of falls and injury. Consider lighting changes that would mitigate these hazards.

SUMMARY

Vision, like all our senses, changes with age. Normal aging can result in reduced vision and a decrease in light sensitivity. In addition, age-related eye diseases can impact clients' ability to live independent and productive lives. Good lighting design can compensate for many of these challenges and allow clients to function conveniently and safely in their living environment. It also plays a vital role in preventing falls and injuries within the home. The designer needs to consider the full range of activities clients will engage in as well as the times of day and patterns of natural light within the living space. When dealing with specific vision challenges, it may be beneficial to collaborate with a medical professional, such as an occupational therapist.

REVIEW QUESTIONS

1. What are two changes that occur in the eye as we age? (See "How Vision Changes with Age" pages 35–36)

2. List three age-related eye diseases that affect vision. (See "How Vision Changes with Age" pages 35–36)

3. What are some important techniques to assist an older client with color recognition? (See "Lighting Solutions for Aging Eyes" pages 36–38)

4. Name three lighting techniques that will enhance safety and convenience for vision-challenged or older clients. (See "Special Considerations for Safety and Convenience" pages 40–41)

Speaking about Lighting

The language we use to speak about lighting influences how we think about it. It also affects conversations with clients. Speaking about lighting *without naming the equipment* helps us to better conceptualize designs and communicate our ideas.

Learning Objective 1: Discuss lighting without naming equipment.

Learning Objective 2: Distinguish different effects of light.

Learning Objective 3: Analyze light in terms of layers.

Learning Objective 4: Distinguish different methods of applying light.

Learning Objective 5: Approach lighting problems using a what/where/how methodology.

LIGHTING EFFECTS, NOT EQUIPMENT

Clients do not want a room simply filled with lighting equipment. They want a beautiful and functional space, such as the kitchen seen in Figure 6.1. They want to be able to see the people, tasks, architecture, and design. They want to be able to change lighting to accommodate different preferences and activities. Sometimes they don't want to see the lighting at all.

When you describe lighting in terms of the equipment, you may visualize the hardware, not the lighting effect. You may limit your imagination and shut out other approaches. You may confuse your clients, who might not recognize the products or understand their effects. Perhaps the biggest risk is failing to understand how the space should appear when it is finally lighted.

This chapter provides vocabulary to speak about lighting without naming the equipment. Here we develop language to:

- Express the overall feeling, the emotion, of a space.
- Describe different effects, or layers, of lighting.
- Analyze lighting as a design problem.

Of course, the specific lighting equipment *is* important—both its technical and aesthetic attributes. We address these issues when we consider how to develop a lighting design in more detail.

FIGURE 6.1 This kitchen includes a variety of light sources including ample daylight.
Design by Carl Bruen, CGR, codesigner Robin Bruen and Debbie Kerr, CKD, Bruen Design Build, Inc., Morristown, NJ
Photo by Wing Wang

Lighting in the Design Process

This book approaches lighting from the perspective of a simplified design process. It uses:

- *Programming* to identify needs, preferences, and constraints.
- *Schematics* to develop a concept for lighting the space.
- *Development* to ensure that the concept meets the programming requirements and to work through the details.
- *Documentation* to communicate the design to the rest of the project team.
- *Management* to resolve the inevitable problems that arise during construction.

This chapter lays the groundwork for programming and schematics.

HOW SHOULD A SPACE FEEL?

How should this kitchen or bath *feel* when it is lighted? What is the overall impression you want the lighting to create?

FIGURE 6.2A This bathroom, with light surfaces and fixtures as well as ample lighting and daylight, is an example of a bathroom considered to be a bright room.

Design by Corey Shannon Klassen, CKD, CBD, codesigners Ian MacDonald and Scott Lumby, Corey Klassen Interior Design, Vancouver, BC
Photo by Jason Karman

Perhaps these are the most important questions to address when you start to think about the lighting. Of course, lighting serves the overall design concept. Lighting is not an end in itself. So these questions really pertain to the space and activities.

Brightness and First Impression

Does anyone want a *dim* kitchen or a dim bath? Probably not. But would everyone say they want a *bright* kitchen or bath? Again, probably not, especially if glary lights come to mind.

It helps to think and speak of the brightness of the kitchen or bath and of the surfaces and objects you see rather than the brightness of the lighting itself. Suggest a range of brightness ("how bright") rather than a bright or dim dichotomy (see Figures 6.2a and 6.2b).

Emotion and Activity

Should the kitchen feel relaxed, homey, social, efficient, clean? These terms are intended to elicit preferences for one *emotional* response or another and often imply different activities.

The words "relaxed," "homey," and "social" suggest that the kitchen also hosts meals and social interaction (see Figure 6.3a). "Efficient" and "clean" suggest a kitchen more devoted

FIGURE 6.2B This bathroom, with dark surfaces and fixtures as well as ample lighting and daylight, is an example of a subdued room.

Design by Holly Rickert, codesigner Julia Kleyman, Ulrich, Inc., Ridgewood, NJ
Photo by Peter Rymwid Architectural Photography

to preparation (see Figure 6.3b). Choose the language you prefer; the idea is to associate evocative terms with lighting effects.

As discussed in Chapter 3, "Seeing the Space and Each Other," lighting that supports relaxed, homey, and social feelings tends to be less uniform (localized brightness) and more peripheral (bright vertical surfaces). Lighting that supports feelings of efficiency and cleanliness tends to be delivered from overhead and to create more uniform brightness. Consider how the lighting arrangements support the overall feeling of the designs.

How do the emotions and activities differ in master, children's, and guest bath areas? What does that suggest in terms of the emotional qualities of the lighting?

FIGURE 6.3A Example of a kitchen designed for social interaction as well as food preparation

Design by Christine Pandur, AKBD, and Tammy MacKay, AKBD Design Eye Ltd., Edmonton, AB
Photo by Merle Prosofsky

FIGURE 6.3B Example of a kitchen designed more for food preparation

Design by Brian M. Johnson, NCARB, Collaborative Design Architects, Billings, MT
Photo by Phil Bell

LAYERING

Designers often speak in terms of layers: layers of clothing, layers of furnishings, and layers of lighting. When you hear the term "layered," you might think of a jacket over a vest over a shirt, or a throw over a pillow, over a sofa, over a rug.

In terms of lighting, layering both describes the luminous experience and helps to analyze its design. The visual experience of layering is more metaphorical than visual: It is more like tasting complex flavors, smelling a blended aroma, and hearing harmonies than seeing a physical or geometric arrangement. The layers of lighting both combine and stand out, creating a subtle and rich visual experience.

Layering also provides a practical approach to analyzing the different effects of a lighting composition. The language of lighting layers helps to deconstruct a lighted space in which you are standing so you can understand how it works. It also helps to construct a lighted environment so that it works as you intend it to.

Richard Kelly (1910–1977) was the lighting designer responsible for some of the most widely used language of lighting and is considered one of the pioneers of architectural lighting design.

Layered lighting effects can be *experiential* . . .

- Focal
- Ambient
- Sparkle

. . . or *functional* (note that some of the terms are similar):

- Task
- Ambient
- Accent
- Wall
- Decorative

We explore both sets of terms because each offers a different but useful approach to seeing and communicating lighting.

Origins

Richard Kelly, the eminent mid-twentieth-century lighting designer, created the descriptive language that underlies so much conversation in lighting practice.

Writing more than 50 years ago, Kelly divided lighting into three primary effects:

1. Focal glow
2. Ambient luminescence
3. Play of brilliants

More plainly, these lighting effects are called focal, ambient, and sparkle, which is the language we use here.

EXPERIENTIAL LAYERS OF LIGHT

Here is how Richard Kelly described the layers of light in "The ABC of Lightplay at Home" (1957): "A is for the Attraction of Focal Glow. B is for the Background of Ambient Luminescence. And C is for the Charm of a Play of Brilliants."

Focal glow	→	Focal light
Ambient luminescence	→	Ambient light
Play of brilliants	→	Sparkle

Note that Kelly describes the light that you *experience*, not the lighting you apply. In the discussion that follows, we begin with Kelly's description and then suggest how the effects can be achieved.

Focal Light

As Figure 6.4 illustrates, a glowing object or surface stands out, attracts your gaze and focuses your attention. Focal light supports the human propensity to look at what is brightest around us.

Objects in museum galleries, retail displays, and theatrical presentations demonstrate the pull of focal glow. Kelly wrote about the effect of a beam of sunlight on a shaded pathway as an example from nature (see Figure 6.5).

FIGURE 6.4 The attraction of focal glow

Courtesy of Tammy MacKay, AKBD, Design Eye Ltd., Edmonton, AB

FIGURE 6.5 Sunlight on a tree-lined path
Courtesy of Creative Commons-Share Alike 3.0, http://www.ForestWander.com

The term "focal glow" describes an object brightened by a beam of light—a vase of flowers, for example—or an object luminous in itself—such as a table lamp with a diffusing shade. Both stand out and draw attention.

The term "focal light" describes the concentrated beam that illuminates the object and creates the glowing effect. Objects need to be three to five times as bright as their surroundings in order to draw attention and perhaps ten times as bright in order to command attention.

Ambient Light

Ambient luminescence—Kelly's phrase—describes the background brightness that orients people to and through a space, offering a sense of security and comfort. Ambient light, as seen in Figure 6.6, fills a space, eliminating the unintentionally dark areas that would (by virtue of their contrast) otherwise call for attention.

The comfort of ambient light arises both from the absence of disturbing and potentially dangerous areas of darkness and also from reducing adaptation to wide variations in brightness.

As an experience, ambient light need not be completely uniform; some variation is almost unavoidable. Consider how difficult it is to achieve uniform brightness in an environment with surfaces of varying reflectance.

More important, perfectly uniform brightness can feel disorienting as it diminishes the spatial and visual hierarchies in the built environment (think of a room with uniformly white surfaces).

FIGURE 6.6 A kitchen filled with ambient light
Design by Christine Pandur, AKBD, codesigner Tammy MacKay, AKBD Design Eye Ltd., Edmonton, AB

Sparkle

What Richard Kelly called "the play of brilliants" we know as sparkle. We recognize sparkle in the crystal pendants of a chandelier, the light glinting in a wine goblet, the twinkling of holiday lights against an evergreen, sunlight reflecting off wavelets in a wind-brushed lake, or light captured by droplets of water after a spring rainfall.

Poetic references to sparkle are appropriate because of the special emotional effect produced by this quality of light. Think of holiday and celebratory experiences, and you frequently see sparkle in the lighting effects (see Figure 6.7). Sparkle is that quality of light that makes you feel good.

As with focal glow, you can create sparkle directly by surrounding small light sources with refractive (light-bending) materials, like the faceted crystal pendants in Figure 6.8.

Or you can create sparkle by directing a beam of light onto materials that disperse the beam, forming the multiple points of light that we call sparkle. Think of the effect of a beam of light on table settings (see Figure 6.9).

Using the Experiential Layers of Light

Kelly taught that lighting experience can be divided into focal, ambient, and sparkle. These experiential layers of light tell us how to look for light in real spaces and also how to imagine lighting spaces not yet built.

Consider the kitchen seen in Figure 6.10. Can you identify each of the layers of light in this room? As an exercise at the end of this chapter, you will have a chance to practice in real spaces.

FIGURE 6.7 Holiday sparkle

Design by James Sasko, Teakwood Builders, Inc., Saratoga Springs, NY
Photo by Scott Bergmann Photography

FIGURE 6.8 A chandelier featuring crystal pendants creates sparkle.

Courtesy of Tammy MacKay, AKBD, Design Eye Ltd., Edmonton, AB

FIGURE 6.9 Sparkle from a table setting
Design by Felicia Gimza, IDDP, CDECA, The Expert Touch Interiors, Oakville, ON

FIGURE 6.10 Kitchen with layered light
Design by Nicholas Geragi, CKD, CBD, codesigner Damani King, Klaff's Inc., Norwalk, CT

FUNCTIONAL LAYERS OF LIGHT

In the preceding section, we discussed experiential layers of light—vocabulary for describing lighting effects. In this section, we develop a slightly different vocabulary, one that describes the functions of light in the space. Both experiential and functional vocabularies will prove helpful in designing light for kitchens and baths.

Functional layers of light are often related quickly to specific luminaires. Unfortunately, this shortcut may make it difficult to envision the lighted quality of the space (as opposed to the equipment in it) and may lead to confusion when trying to develop lighting from concept to a fully specified design.

We begin this discussion with task light because reflections from task surfaces contribute to the overall (ambient) brightness of the space. Experience shows that designing for *task and then for ambient light* tends to create layers of light that more closely conform to recommended practice in task-oriented spaces, such as kitchens and baths.

Task Lighting

The term "task lighting" refers to lighting directed principally to illuminate work surfaces and help people perform tasks (see Figure 6.11). In kitchens and baths, the primary task areas in most homes, task lighting is particularly important.

Recommendations for the quantity of task illumination refer to the illuminance on the task—regardless of whether that illumination is produced by a dedicated task-oriented luminaire or by other lighting equipment providing ambient light.

In some applications, such as laundry rooms, task lighting may be delivered entirely from what might be considered the ambient layer, as seen in Figure 6.12.

Ambient Lighting

Ambient lighting provides the overall illumination that eliminates unwanted areas of darkness and balances the brightness of the task and accent layers discussed next.

Since ambient lighting covers many of the work surfaces in a space, it contributes to task illumination. In kitchens, overhead ambient lighting also provides the illumination for the visual tasks inside upper cabinets (see Figure 6.13) and for cleaning the floor.

Accent Lighting

Accent lighting—like focal light—illuminates objects to draw attention to them (see Figure 6.14). As we noted earlier, in order for lighting to accent or highlight an object, it must increase its brightness significantly—three to five times—compared to the surroundings. Accent lighting also can be used to highlight architectural details of a space. An example may be cove ceilings or tray ceilings. In this case, the brightness does not have to be three to five times brighter than the rest of the space. This type of accent lighting results in a nice aesthetic and brings attention to architectural types of details.

Wall Lighting

As we discussed in Chapter 3, lighting that illuminates the periphery of a space significantly affects our sense of spaciousness and relaxation. So it makes sense to devote a separate layer of lighting to the periphery.

Busy spaces, such as kitchens, may not expose any walls; cabinetry and appliances fill most of those surfaces. Ambient lighting provides the layer that illuminates the exteriors of most cabinetry. Illuminating the inside of a glass-front cabinet is often considered an accent layer (Figure 6.15).

FIGURE 6.11 Note the task lighting designed for this kitchen to meet the challenge of the high vaulted ceiling.

Design by David McFadden, codesigner Debbie Larson, Past Basket Design, Geneva, IL Photo by David McFadden

Nevertheless, the wall layer reminds us of the importance of this peripheral lighting function in the lighting composition.

Decorative Lighting

In an arrangement of otherwise functional layers, decorative lighting provides light from lighting equipment that offers a dominant glow or sparkle effect.

Some applications, such as a guest bath, often use decorative wall sconces or pendant lighting to establish a design theme or simply to impress visitors (Figure 6.16).

FIGURE 6.12 Laundry room with ambient light

Design by Divine Custom Homes, Hudson, WI
Photo by Spacecrafting

FIGURE 6.13 Ambient lighting is provided by the recessed lights in this ceiling design while chandeliers above the islands provide additional task lighting.

Design by Sandra L. Steiner-Houck, CKD, Steiner & Houch, Inc., Columbia, PA
Photo by Peter Leach

FIGURE 6.14 Accent lighting illuminates the objects displayed in this lavish bathroom.

Design by Brigitte Fabi, CMKBD, Drury Design Kitchen & Bath Studio, Glen Ellyn, IL
Photo by Eric Hausman

FIGURE 6.15 The lighting inside the glass cabinets draws attention to the glassware on display.

Design by Lauren Levant Bland Jennifer Gilmer Kitchen & Bath LTD. Chevy Chase, MD
Photo by Bob Narod, Photographer, LLC

FIGURE 6.16 A large decorative pendant with a clear glass shade lights this bathroom featuring a variety of wall and ceiling surface materials.

Design by Tammy MacKay, AKBD, Design Eye Ltd., Edmonton, AB

Decorative lighting also can be functional, such as glass pendants that provide task lighting for a kitchen island (see Figure 6.17).

Using the Functional Layers of Light

The functional layers of light guide us in thinking about what lighting should *do*: illuminate tasks, provide comfortable ambient illumination, accent special objects, brighten walls, and decorate a space with luminous objects.

In this sense, the functional layers are analytical; they can be recognized most easily by identifying the luminaires associated with each layer. That is both the strength and weakness of this approach. Using Figure 6.18, identify the functional layers of light.

FIGURE 6.17 Pendants provide task lighting over the island in this kitchen featuring a vibrant color scheme.
Design by Cheryl Kees Clendenon, codesigner Stacy Snowden, In Detail Interiors, Pensacola, FL
Photo by Greg Riegler Photography

Functional or Experiential Layering of Light

The functional and experiential layering of light is similar but not identical.

As an exercise, associate each of the five functional layers with one of the three experiential layers. Now explain how they differ from each other.

APPLYING LIGHT

When we say, "Let's use recessed lighting," that is shorthand for light directed down from the ceiling. But it does not indicate whether the light will flow in a concentrated beam or a more diffuse one or even asymmetrically. And, depending on the listener's experience, the phrase "recessed lighting" may conjure small, sparkling luminaires, large cumbersome ones, or something altogether different.

Regardless, at this stage of the design, the phrase "recessed lighting" is too vague to describe the effect of the light and perhaps too specific (in the mind of the listener) to describe the luminaire itself.

FIGURE 6.18 This lavish bathroom design features a variety of lighting sources.
Design by Sandra L. Steiner-Houck, CKD
Photo by Peter Leach Photography

The next section provides vocabulary to communicate how light can be applied in a space—that is, how the various layers of light can be achieved.

Direction

We commonly speak of the direction of light in spatial terms: down light and up light. While these terms are unambiguous, they do not express the *effect* of the light.

Effect is better conveyed by how the light reaches a target.

- *Direct light* reaches a target directly, without reflecting off another surface. Since direct light is not diffused by reflection, it tends to be more concentrated and so produces more highlight and shadow. Direct light does not imply down light, although that is the most common example. A table lamp or an under-cabinet luminaire produces direct lighting effects (see Figure 6.19).

FIGURE 6.19 The under-cabinet lighting below the microwave and the table lamp in the adjoining room are examples of direct light.

Design by Mark T. White, CKD, CBD, Kitchen Encounters, Annapolis, MD
Photo by Phoenix Photography

- *Indirect light* reflects off a surface (typically the ceiling or a wall). Reflecting off a matte surface diffuses the light, which spreads out and diminishes the form and texture of the objects it ultimately illuminates (see Figure 6.20).
- *Backlight* reaches an object from behind and does not reflect in the direction of the viewer. As a result, the object becomes silhouetted against the source of the light (see Figure 6.21).

Note that light reflecting off a *specular* surface (such as a mirror) does not diffuse and acts more like the direct light from the source.

FIGURE 6.20 The lighting fixtures in this bathroom are examples of indirect light.

Design by Lori W. Carroll, ASID, IIDA, codesigners Debra Gelety, Allied ASID, EDAC, and Mary M. Roles, Lori W. Carroll & Associates, Tucson, AZ Photo by William Lesch Photography

Concentration

As noted, *concentrated* light produces sharper highlights and shadows, revealing form and texture. *Diffuse* light softens shadows and diminishes the perception of form and texture.

FIGURE 6.21 Backlighting was used for this striking wall design behind the cooktop.

Design by Kaye Hathaway, CKD, NCIDQ, ASID, codesigner Catherine Heir, Dea Design Group, LTD, Island Lake, IL
Photo by Jozef Jurcisin

Orientation

Now it is appropriate to speak of downlight and consider whether the light is oriented to a wall or vertical surface. Table 6.1 combines the effects of direction, concentration, and orientation, showing typical effects produced.

What, Where, How

Another analytical approach to lighting is to ask three fundamental questions:

1. What should be lighted?
2. How should it appear when lighted?
3. Where can it be lighted from?

TABLE 6.1 Effects of direct and indirect lighting

Direction	Concentration	Orientation	Effect
Direct	Concentrated	Down	Focal pool of light
	Concentrated	Wall	Focal accent on painting
	Concentrated	Down the wall	Focal graze
	Diffused	Down	Ambient light
	Diffused	Wall	Ambient wall wash
Indirect	Diffused	Up	Ambient light
	Diffused	Wall	Backlight and silhouette

TABLE 6.2 Lighting recommendations

What	Brightness	Color Rendering	Distribution
Counter tasks	High	Excellent	Diffused to avoid shadows
Cooktop tasks	High	Excellent	Concentrated on task
Cabinet tasks	Medium	Good	Diffused over cabinet fronts
Overall space	Medium to low	Excellent	Diffused to avoid shadows
Collectibles	Medium	Good	Focused on objects
Paintings on wall	Medium	Excellent	Diffused to cover all paintings

Clearly, these questions do not specifically address layers of light. Instead, they organize the design process.

What Should Be Lighted?

Applying the functional layers of light, our list might include:

- Counter tasks
- Cooktop tasks
- Cabinet tasks
- The entire room
- Collectibles in the breakfront
- Paintings on the wall

See Table 6.2 for lighting recommendations.

How Should It Be Lighted?

Chapters 2 to 4 discussed looking at materials, spaces, and tasks. What characteristics would we use to describe the desired lighting effect? Brightness, color, and the distribution of light (concentrated or diffused) would certainly rank as three of the most important considerations.

Table 6.3 is a simplified approach to answering the question, "How should it be lighted?" At this point, specific illuminance levels are not part of the design process; however, relative brightness is important.

Arguably, all of the lighting should deliver excellent color. And a single level of color quality is often used. Nevertheless, color quality is more significant when looking at people and art than when doing something like reading, where color is not an important factor. In chapters 10 and 11 we define color quality in terms of light source attributes (color temperature and color rendering index).

TABLE 6.3 Recommended Locations for Lighting

What	Issues	Location
Counter tasks	Shadows created by users Veiling reflections	Under cabinet At front edge of cabinet
Cooktop tasks	Shadows created by hood	Integrated in hood
Cabinet tasks	Reaching top shelves Avoiding sharp shadows on door	Ceiling, moderately close to cabinets
Overall space	Nondistracting diffuse quality	Ceiling
Collectibles	Clear rendition of form	Inside breakfront, at front edge
Paintings on wall	Several small paintings	Ceiling, parallel to wall

Where Should It Be Lighted From?

Sometimes architecture determines the location of the lighting equipment. Sometimes location depends on the quality of light desired. Although the ceiling provides a convenient location for lighting equipment, both people and cabinetry can get in the way.

The locations for lighting in Table 6.3 are typical but not universally recommended. For example, high or sloped ceilings often make recessing lighting equipment impractical. We look at the location of luminaires in detail when we cover design development in Chapter 14.

Looking at Layered Lighting

Use your lighting journal to record your observations in this exercise. You can do this in your own kitchen or bath, in other rooms in your residence, or at work.

1. Make a sketch of the space so you can easily identify lighting.
2. Deconstruct the lighting in the space into the three experiential layers of light.
3. Note where you experience focal glow. Does this glow provide appropriate focus to your experience of the space?
4. Note where you experience sparkle. What does the sparkle contribute to your experience of the space?
5. Characterize the ambient light as to its brightness and uniformity, looking at all surfaces that contribute.
6. Next, reanalyze the lighting in terms of the five functional layers of light. You may find it helpful to make a second sketch of the space.
7. How does this analysis differ from the first one you did?
8. Which analysis better describes how a visitor would see the lighting? Why?

SUMMARY

Layered lighting can be considered both experientially and functionally. Experiential layers include focal, ambient, and sparkle. Together they help you see a lighted space and provide a visual vocabulary for communicating lighting ideas.

Functional layers include task, ambient, accent, wall, and decorative. They help you analyze lighting against a client's design program.

Asking what, where, and how can help you put layered ideas into a design process.

REVIEW QUESTIONS

1. What are the three effects of lighting proposed by Richard Kelly? (See "Experiential Layers of Light" page 48.)
2. Discuss the difference between focal glow and focal light? (See "Focal Light" page 49.)
3. List and define the five functional layers of lighting. (See "Functional Layers of Light" page 54.)
4. What are the principal differences between the experiential and functional layers of light—do *not* use those words in your explanation. (See "Experiential Layers of Light" page 48, and "Functional Layers of Light" page 54.)
5. How do the questions What?, Where?, and How? relate to the functional layers of light? (See "What Should Be Lighted?" page 64.)

Sustainable Lighting

Sustainable lighting applies the principles of resource usage over time to lighting choices. This chapter first defines sustainable lighting. We then look at lighting to discuss its specific environmental impacts and compare its financial and environmental costs. Finally, we look at broad strategies for building a sustainable foundation for lighting design.

Learning Objective 1: Recognize the environmental impacts of lighting.

Learning Objective 2: Explain the costs of lighting as related to sustainability.

Learning Objective 3: Consider lighting in a sustainability context.

DEFINITION OF SUSTAINABILITY

In 1987, the Brundtland Commission of the United Nations wrote what has become the most widely used definition of the term "sustainability": "Sustainable development is development meeting the needs of the present without compromising the ability of future generations to meet their own needs."

Over 25 years, the concept of sustainability has undergone challenge and refinement, influenced the thinking of professionals and consumers, and stimulated industrial activity.

The result of this refinement has led to the commission establishing the familiar three pillars of sustainable development: economic growth, environmental protection, and social equality.

When it comes to making lighting choices for a specific project, a more practical definition of lighting sustainability might be: "Sustainable lighting meets human needs with the least impact on the natural environment."

Thus, the prerequisite for sustainable lighting is meeting human needs. Lighting that fails to provide adequate illumination for safe passage, task performance, and overall well-being is inherently wasteful and therefore is not sustainable, regardless of how little carbon is released into the atmosphere.

What about lighting for beauty and image? A beautiful kitchen may not be at the top of the hierarchy of human needs, but it still serves an important human desire, one that is inextricably tied to our social relations.

FIGURE 7.1 This award-winning kitchen incorporates many sustainable design solutions.
Design by Jessica Williamson Kitchen Views, Newton, MA
Photo by Michael J. Lee Photography

Meeting human needs is necessary—but not sufficient—for sustainability. Sustainable lighting must meet human needs with the least impact on the natural environment (see Figure 7.1).

ENVIRONMENTAL IMPACTS OF LIGHTING

Indoor lighting, such as lighting in kitchens and baths, affects the natural environment in three principal ways:

1. Impact of electricity consumption
2. Quantity of material used initially and over time
3. Potentially harmful material content

Electricity: Generation and Consumption

From the beauty of a glowing source of light to the physics of electricity—here we go.

Electricity is produced mostly by spinning an electromagnetic generator. As the magnetic components rotate, they induce electric current, which is transmitted at high voltage. To rotate the electromagnet, electric utilities heat water to create steam, which operates on the blades of a turbine, changing the form of energy from thermal, to rotational (kinetic), to electric.

To create the steam, utilities burn coal, natural gas, or oil, all of which emit carbon (and other contaminants) into the atmosphere. Carbon emissions, in the form of carbon dioxide (CO_2), trap heat in the atmosphere, much as a glass-enclosed greenhouse does. Hence, the term "greenhouse gas" to describe atmospheric CO_2 and other heat-trapping gases.

Despite the political and economic debates around global climate change, the science is pretty clear. Greenhouse gas accumulation poses serious, potentially catastrophic threats to our way of life. Here is the irony: It is a way of life powered in large part by electricity.

Carbon emissions depend significantly on the fuel used and how it is burned. Generally speaking, coal causes significantly more problems than natural gas. (Of course, so-called clean sources—nuclear, wind, and water—present problems of their own.) Regardless of the fossil fuel source, the more electricity is generated, the more carbon is emitted.

 Three factors affect how much electricity will be used:

1. The quantity of light you apply
2. The amount of time you use it
3. How efficiently you convert electric energy to visual energy

We return to these factors when we discuss sustainable lighting strategies later in this chapter.

Lighting consumes a great deal of the electricity we generate and use—more than 15 percent of electricity used in residences, and more still in commercial applications. Electricity itself is not the problem. Indeed, electric lighting is significantly cleaner and safer than gas lighting, oil lamps, candles, wood fires, and other primitive forms of human-made illumination.

It is the emissions from electricity generation plants that affect the environment (see Figure 7.2).

FIGURE 7.2 Electric power plant emissions

Embedded Energy

When discussing sustainable lighting, embedded energy should also be considered. Embedded energy refers to the total energy required in the manufacturing of a product as well as the transportation of the product to the final site where it will be used. Energy is embedded in the production of lighting equipment and the processing of key components, such as aluminum and glass. Once the equipment is produced, it must be transported from the factory to the building site (typically in at least two separate trips). Today, much of the lighting equipment used residentially in North America is manufactured, all or in part, in Asia.

Material Usage

Since all resources ultimately are finite, material usage has an important environmental impact.

Throughout history, civilizations that failed to manage resource usage have suffered, some to extinction. Management, of course, includes conservation, advanced exploitation, and technological innovation, but also relocation and conquest.

Usage combines the quantity of resources used initially and how much needs to be used over time. Imagine you are holding two light sources for typical household use: a light-emitting diode (LED) in your left and a compact fluorescent (CFL) in your right (see Figure 7.3). Both are designed to fit into a typical screw-shell socket. The LED weighs about 178 grams; the CFL weighs about 153 grams. Which uses more material? That is easy . . . or is it?

The LED is rated at 25,000 hours, and the CFL is rated at 10,000 hours. So you will use almost three of these CFLs to match the rated life of the LED. Considering the number of CFLs used in that period of time (about 6 years if lights are on 12 hours per day), the combined weight

FIGURE 7.3 LED and CFL lamps
Creative Commons Share Alike 3.0

of CFLs would exceed 450 grams (3 × 153 = 459). That's actually two and half times the weight of the LED.[*]

Let's do a little more math. If lights are on 12 hours a day (all day or all night), the annual usage is 12 × 365 = 4380 hours per year. Divide the life of the LED source by the annual usage: 25,000 ÷ 4380 = 5.7 years.

So, to accurately compare material usage for lighting equipment, you need to consider not only the amount of material with respect to manufacturing the source initially, but also how long it will last and the landfilling at the end of its life.

Material Content

Overwhelmingly, lighting equipment has been manufactured from nonrenewable materials: metal, glass, plastic, ceramic, and various minerals. Structural, optical, conductive, and insulating requirements demand that certain materials are used. In some instances, a recycled version may be used. Ideally this is preferred from an environmental standpoint; however, not all materials are available in a recycled form.

Most manufacturers recycle waste material from their fabrication processes. Cost reduction rather than environmental concerns typically drives this "preconsumer" recycling.

Some materials with substantial recycled content, such as aluminum for extrusions and castings and cardboard for packaging, are commonly available to manufacturers. However, there are no standards for the amount of material that must be recycled.

Does lighting rely on dangerous materials?

Applying a strict filter, all popular electric light sources—incandescent, fluorescent, and LED—do contain very small amounts of potentially hazardous materials. But, despite occasional and sensational news stories, lighting equipment has proved itself quite safe, with most problems arising from fire, not material contamination. Nevertheless, mercury, which is an essential ingredient in energy-efficient fluorescent (and other) lamps, stirs considerable debate.

Although client interest may focus on material content, particularly mercury, it is worth remembering that the dominant environmental impact of lighting is energy usage, which depends on the amount of lighting used, for how many hours, and the efficiency at which electricity is converted to light.

Life Cycle Assessment

The *Life-Cycle Assessment of Energy and Environmental Impacts of LED Lighting Products* was published in three parts in 2012 and 2013. At the time of this writing, these assessments present the most comprehensive analyses of LED products in terms of energy and environmental impact.

Part 1 covers energy usage, Part 2 covers LED manufacturing and performance, and Part 3 covers LED environmental testing.

Quoting from Part 2, "Overall, this study confirmed that energy-in-use is the dominant environmental impact, with the 15-watt (W) CFL and 12.5W LED lamps performing better than the 60-W incandescent lamp."

(continued)

---

[*] The weights of the two lamps (LED versus CFL) are from *Life-Cycle Assessment of Energy and Environmental Impacts of LED Lighting Products, Part 2,* available free online from the US Department of Energy (DOE) (http://apps1.eere.energy.gov/buildings/publications/pdfs/ssl/2012_led_lca-pt2.pdf).

> Regarding potentially toxic elements, the Assessment Summary states:
>
> The selected models were generally found to be below restrictions for Federally regulated elements.
>
> Nearly all of the lamps (regardless of technology) exceeded at least one California restriction—typically for copper, zinc, antimony, or nickel.
>
> Examination of the components in the lamps that exceeded these thresholds revealed that the greatest contributors were the screw bases, drivers, ballasts, and wires or filaments.
>
> Concentrations in the LED lamps were comparable to concentrations in cell phones and other types of electronic devices, and usually came from components other than the LEDs themselves.

Mercury

Mercury is a neurotoxin, with the potential to cause severe developmental disabilities. Mercury poisoning most often travels up the food chain, following contamination of water and ingestion by fish. Mercury vapor can also be released by burning coal.

Forty years ago, when the mercury content in fluorescent lamps was five to six times that of today's lamps, there was little outcry, despite longtime awareness of mercury's danger and the general use of mercury-containing fluorescent lighting in schools.

Beginning in the 1980s with efforts to improve energy efficiency overall and promotion of CFL lamps in homes, mercury concerns spread.

Disposal of fluorescent lamps in municipal landfills, with the potential for mercury to leach out into the water system, was a key concern. Today's fluorescent lamps generally have much lower levels of mercury than earlier versions (as much as 90 percent lower). Additionally, many lamps are tested to guard against mercury leaching out into landfills.

Finally, despite alarmist media reports about mercury hazards, you might worry more about a glass cut on your foot from a broken fluorescent lamp than the risk of mercury contamination in your home.

LED lighting contains *no* mercury, which may affect buying decisions.

Mercury Testing

There are several methods of assessing mercury content in fluorescent lamps. The most commonly used method is the US Environmental Protection Agency Toxic Characteristic Leaching Procedure (TCLP), which endeavors to simulate what might happen in a landfill. To perform the test, the lamp is disassembled, ground up, and analyzed to determine whether potentially toxic elements can leach out. Lamps that pass the TCLP test are commonly marketed as low mercury.

Recycling

Simple lighting equipment recycles easily along with other construction waste. Electronics, however, pose some challenges in segregation. Some manufacturers of LED luminaires promote their products as recyclable, although no standard backs up this claim.

Recycling spent light sources is subject to *state* regulation, which varies by jurisdiction. Federal regulations designate fluorescent lamps as hazardous waste, but this alone does not mandate recycling.

Today, the cost of recycling exceeds the value of recovered materials, so commercial lighting users must pay to recycle spent fluorescent and other glass-based lamps. Residential users often can recycle CFL lamps at home center stores (see Figure 7.4) at no charge. (The cost, which is real, is absorbed by the stores.) Linear fluorescents, however, are not acceptable at these sites.

FIGURE 7.4 Consumer recycling bins
Courtesy of The Home Depot

Recycling separates and recovers mercury from spent fluorescent lamps, and the mercury then is reused in new fluorescent lamps. Recycled glass lacks the optical purity to be reused and is often recycled into lower-grade building material, such as masonry blocks.

Recycling of LED lighting equipment appears promising. (No regulations currently exist.) Because LED lighting contains significant amounts of aluminum and other materials, the opportunity for financial recovery and a market-based solution exists.

SUSTAINABILITY AND THE COST OF LIGHTING

Mention the cost of lighting casually, and most people promptly think of the price of the lighting equipment. That is a simple number, but it is neither a full nor a meaningful picture.

Considered broadly, lighting presents two costs: the financial one you pay and the environmental one you do not. More accurately, you do not pay the environmental cost *immediately*; you share it with neighbors and succeeding generations.

We looked at the environment costs—impacts—of lighting in the previous section. Now we turn to the financial cost, which breaks into four characteristic components:

1. Material (luminaires, controls, wiring)
2. Electricity (to operate the lighting)
3. Labor (to install and maintain the equipment)
4. Disposal (to recycle spent or replaced equipment)

This approach yields the convenient mnemonic M-E-L-D. You can also consider cost as initial cost (material and installation labor) and operating cost (electricity, replacement material, service labor, and disposal).

We return to cost in more detail when we discuss budgeting, design development, and construction. Now we take a broad look at the relative impact of the different financial cost components using a sustainability perspective.

Material Cost

Which has a higher material cost: a 65-W incandescent reflector lamp priced at $5.00 or a 13-W LED reflector lamp priced at $35?

By now you recognize that the *life* of the lighting equipment affects its real cost (as it affects its environmental impact).

In this example, the incandescent lamp has a rated life of 2000 hours and the LED lamp has a rated life of 25,000 hours, which is 12.5 times as long. To operate as long as the LED lamp, you would use 12.5 incandescent lamps, at a cost (rounded down) of $60.

Time Value of Money

This TCO analysis ignores the time value of money. Costs today are valued the same as costs in the future, with adjustment for interest rates. For financially sophisticated readers, this may be a serious oversimplification. However, it makes the calculations much simpler.

Two considerations mitigate this problem:

1. Energy costs dominate the analysis, and the electricity rate is flat (no inflation) over the calculation period.
2. Shortening the life cycle analysis period is a reasonable approach to introducing time value.

Electricity Cost

The formula for calculating electricity cost is straightforward:

$$\text{Watts} \times \text{Hours} \times \text{Rate}$$

- *Watts* are the load of the lighting system, including all of its components .
- *Hours* are the hours that the lighting operates. The DOE uses 3 hours per day, 1095 hours per year, as the residential average. We typically use 6 hours per day (2000 hours per year) for the longer operating time in kitchens and baths.
- *Rate* is the effective cost of electricity per kWh of use. A kWh is 1000 watt-hours. The current average used by the DOE is $0.11 per KWH, which we use, unless otherwise noted.

Let's replicate the electricity cost calculation for the 60-W lamp. First, set up the equation using the data given.

Now you can resolve the formula with simple arithmetic. 60W must be converted to kilowatts:

$$\frac{60\,W \times 1\,KW = 0.06\,KW}{1000\,W}$$

2000 HRS (Kitchen & Bath Average/Year)

Rate: $0.11/KWH

So:

Cost = 0.06 KW × 2000 HRS × 0.11 KWH

Cost = 120 KWH × 0.11 KWH

Cost = $13.20/year

A 60 watt lamp costs $13.20 per year to run for 2000 hours.

Two points to note:

1. The financial cost of electricity—like its environmental impact—depends on the power used by the lighting equipment and length of time it operates.
2. The calculations are simpler when you think of time in thousands of hours.

Electricity Pricing

Have you looked at your electric bill lately? It can be pretty complicated. You may find charges for connection, consumption, transmission, services, taxes, fees, maybe more.

If you focus on the consumption charge in dollars per kWh, you may seriously underestimate the real cost of electricity. A better way is to look for the total charge for electricity at the bottom of the bill. Then find the total kWh used. Now divide the dollar charge by the kWh usage to find your effective rate. Just to satisfy yourself, compare the effective rate to the simple consumption rate.

Labor Cost

Installation cost depends on the difficulty of installation (hours per luminaire), the extent of the installation (number of luminaires), and the labor rate (dollars per hour).

These factors are largely independent of the energy consumption or life of the equipment. Nevertheless, advanced energy-saving lighting—especially controls—may challenge installers unfamiliar with the technology. Once energy saving control technology becomes more familiar, installers of this new lighting will find that the simplicity of installation methods has also come a long way.

Maintenance cost depends on the difficulty, extent, labor rate, and the frequency of maintenance. Frequency depends on the life of the equipment to be maintained.

Fortunately, most energy-efficient lighting equipment also enjoys relatively long operational life, which reduces its maintenance cost.

Disposal Cost

For most of us, disposing of spent lighting equipment costs little directly out of pocket. The financial cost is socialized in the cost of municipal waste facilities (and the taxes and fees imposed to support them).

Commercial facilities typically pay for the recycling of spent fluorescent lamps. With large-scale, scheduled maintenance, recycling costs about $0.15 per foot for fluorescent lamps. Compared to the cost of electricity and material for such lamps, recycling costs less than 1 percent of the initial price.

Intermittent maintenance significantly increases the cost of recycling. Even so, recycling remains an insignificant cost of ownership when compared to storage, handling, and transportation costs. Compact fluorescent lamps often are recycled in buckets; the recycling cost per lamp is higher than with linear lamps but still near 5 percent of total cost.

Cost Model

Over the next several pages, we discuss the total cost of ownership (TCO) as it pertains to lighting. It is worth noting what is *not* in this model.

TCO includes only financial costs. It does not include any nonfinancial environmental impacts, nor does it include any costs (or benefits) associated with lighting *performance* (e.g., injuries or the impact of task performance).

Total Cost of Ownership

The total cost of ownership (TCO) assesses the four cost components *over the useful life* of the lighting system. It is a form of life cycle analysis and is especially useful when lighting systems are compared on a comparable basis.

Let's begin with a simple exercise: How much does a typical household lightbulb cost?

First, some data (all from the package on the shelf):

Price at the store:	$0.50 per lamp
Wattage:	60 W
Life:	1000 hours

As long as the bulb sits on the shelf, its cost is just $0.25. As we *use* it, however, the cost of ownership rises with the electricity consumption.

Over a 1000-hour life, 60 W of power consume 60 kilowatt-hours (kWh) of electricity at a cost of $6.60, using US DOE data. Material and electricity costs for this simplest of light sources total $7.10, with electricity accounting for over 90 percent of that amount.

Life Cycle Cost

Life cycle analysis combines installed cost (sometimes called initial cost) and operating cost (i.e., the total of material, electricity, labor, and disposal costs) into a consistent analytical framework. There are three key parameters to consider:

1. *Time frame.* The cost calculation might use the projected life of the structure, occupancy, equipment, or any other period that can is relevant and can be applied consistently.
2. *Performance.* The cost calculation should be based on lighting that meets all relevant project requirements, such as illuminance, color, comfort, and control capability.
3. *Relevant scope.* The calculation might consider the entire installation (if alternative systems are being contemplated). Or it may focus on a single unit of equipment (if comparisons can be made on that basis).

SUSTAINABLE LIGHTING STRATEGIES

The *environmental* impact of lighting consists of three basic components:

1. Emissions from electricity generation powered by fossil fuels
2. Material used (and disposed) in the lighting equipment
3. Any potentially hazardous content in that material

- The *financial cost* of lighting includes both installed and operating cost and includes material, electricity, labor, and disposal.
- *Electricity* represents the single largest environmental and financial cost of lighting.
- The environmental and financial cost of electricity *both* depend on the electrical power in the lighting and the hours it is used.

These points lead directly to practical strategies for sustainable lighting. But first let us consider the client's perspective.

Motivation

Clients use sustainable lighting for three reasons: They value the environment; they expect financial savings; and they have no choice due to regulations. In simple terms, that is *want to*, *pays to*, and *got to*.

- Some clients will ask (even demand) that your design minimize environmental impact as a *priority*. That is *want to*.
- Other clients only wish to adopt cost-saving technologies. If those technologies also reduce environmental impact, then so much the better. That is *pays to*.
- Regardless of your client's wishes, your design will conform to applicable codes, even when those codes add cost. That is *got to*.

Other client—perhaps most—look for a reasonable balance, lighting that both reduces environmental impact and lowers financial cost.

The strategies you adopt for your design ultimately will depend on your client's values. Those strategies include energy efficiency, long equipment life, and clean materials content.

Energy Efficiency

Because energy usage—electricity consumption—imposes the highest cost on the environment, *energy efficiency is the most important strategy for sustainable lighting*. When it comes to lighting, energy efficiency is more important than material usage or content—by far. As a sustainable strategy, energy-efficient lighting applies three basic techniques.

The strategy begins by *using only what you need*. That is, the goal is to minimize the use of light when tasks do not require it, when daylight provides the necessary illumination, or when no one is around to use it. Because you are only preventing waste, this aspect of energy efficiency does not limit lighting's ability to meet client needs.

The following design approaches are ways to achieve energy efficient lighting.

- A layered lighting design with task-specific illumination uses less electricity than a uniform level of illumination.
- Windows provide the critical amenity of view; they also can reduce lighting energy consumption. To accomplish reduced lighting energy consumption, the lighting design combines daylight with controls to reduce illumination when daylight is present.
- Lighting controls can reduce electricity use still further: Switching lights off when no one is present or dimming them when users or occasional tasks require less than full output saves clients' money by reducing their electric bills.
- These controls can be completely programmable with users able to manage the programming from remote devices.

The second technique is *using reflective finishes*. Unwanted glare, however, must be a consideration when using this technique. When light strikes a surface, some is absorbed. The lower the reflectance of the surface, the higher the absorption of light, and the less light ultimately reaches its target. Dark stone and highly textured surfaces, for example, can be light traps. Since these low-reflectance materials can be an important part of the interior design, you face a potential trade-off between visual and environmental impact. Careful design—limiting the use of dark finishes and locating them so they are not important reflecting surfaces—can address the issue (see Figure 7.5).

The third technique—perhaps the most familiar—is *using efficient lighting equipment*: light sources and luminaires. The measure used to evaluate lighting equipment is called luminous

FIGURE 7.5 Careful attention must be paid to the lighting design in rooms that include dark and textured surfaces.
Design by Holly Rickert; codesigner Julia Kleyman, Ulrich Inc., Ridgewood, NJ
Photo by Peter Rymwid Architectural Photography

efficacy—the amount of light produced per unit of power, technically, lumens per watt. Lumens per watt sometimes are shown as lm/W although LPW is widely used in the industry when referring to lumens per watt.

Light sources, of course, differ considerably in their ability to convert electricity into light. Similar-looking fixtures may perform quite differently. The better one perhaps emits more light than the other. However, some efficient fixtures are simply glary. And some efficient light sources distribute light inappropriately or with poor color. It is important to establish lighting needs—quantity and quality—so that the pursuit of efficiency does not come at the expense of lighting needs.

Long Equipment Life

Reducing the *number* of luminaires in order to reduce material consumption is not generally a satisfactory approach to sustainable lighting. Remember, the first priority is meeting lighting needs. The luminaire location contributes directly to the effectiveness of the lighting design. Spacing luminaires farther apart may leave undesirable dark areas, while using fewer brighter luminaires often produces glare.

Using *long-life light sources* is a good sustainable lighting strategy, provided the source provides the lighting needed. Since today's energy-efficient light sources—LED and fluorescent—also enjoy relatively long life, this strategy dovetails well with using energy-efficient lighting equipment.

Clean Materials

Does using a mercury-free light source make lighting more sustainable? Yes, *provided* that the lighting is energy efficient and long lasting. On this basis, LED is a more sustainable choice than compact fluorescent in most cases. But compact fluorescent is substantially more sustainable than incandescent.

Sustainability and Cost

By now, the direct association of environmental impact and financial cost should be clear. On a life cycle basis, sustainable lighting—efficient, long lasting, and clean—costs less than lighting with a heavier environmental footprint.

SUMMARY

The environmental impacts of lighting include the emissions from electric power plants fired by fossil fuels, material usage, and potentially toxic elements in the materials. Sustainable lighting is energy efficient, lasts a long time, and uses clean materials. The financial cost of lighting (TCO) includes initial and operating costs, comprising material, electricity, labor, and disposal. As with environmental impact, energy dominates financial cost.

How Sustainable Is My Space?

Use your lighting journal to record your observations in this exercise. You can do this in your own kitchen or bath, in other rooms in your residence, or at work.

1. Without doing any analysis, how energy efficient do you think your space is?
2. Identify the lighting sources used in the space.
3. Create a table showing, for each type of luminaire, the number, source, and wattage. (You may have to remove the lamp to see the wattage listed on it.)

(*continued*)

4. As a broad guide, list the efficacy (lm/W or LPW) values based on the source. You can source the efficacy by finding the specification of the lamp online at the manufacturer's website.
5. Calculate the weighted average efficacy (LPW) by (a) multiplying the LPW for that source by the wattage, (b) adding all of the (a) values together, and (c) dividing the sum (b) by the total wattage.
6. Put an asterisk by each luminaire type that is controlled by a dimmer.
7. Now that you have analyzed the lighting, how would you rate its sustainability?

REVIEW QUESTIONS

1. Discuss how the definition of sustainable lighting in this chapter compares to the Brundtland Commission definition. (See "Definition of Sustainability" page 67, and "Environmental Impacts of Lighting" page 68)
2. What are the three key environmental impacts of lighting? (See "Environmental Impacts of Lighting" page 68)
3. Compare the environmental impact of LED and compact fluorescent lighting. (See "Electricity: Generation and Consumption," page 68, "Material Usage" page 70, "Life Cycle Assessment" page 71)
4. Discuss four strategies for improving the energy efficiency of a lighting design. (See "Energy Efficiency" page 77)
5. How is the energy efficiency of lighting equipment measured? (See "Energy Efficiency" page 77)

The Importance of Daylight

Daylight is our most powerful, colorful, and dynamic source of light. At peak times, exterior daylight can provide more than 500 times the typical indoor illumination in homes. The spectrum of daylight is full of color—that is, it contains all wavelengths, in relative balance. And daylight changes all the time. In addition to abundant illumination, daylight influences many different aspects of human experience. How well we enjoy the benefits of daylight depends on how we admit daylight into homes, control it, and integrate it with electric light.

Learning Objective 1: Describe how daylight affects the human experience.

Learning Objective 2: Analyze the forms and characteristics of daylight.

Learning Objective 3: Consider how to admit daylight into the interior.

Learning Objective 4: Integrate daylight and electric lighting.

DAYLIGHT AND HUMAN EXPERIENCE

Daylight is humanity's first and perhaps most influential experience of lighting.

Walk under the shade of a generously canopied tree and enjoy the dappled light through its branches (see Figure 8.1). Venture beyond the reach of the leaves and feel the beat of the summer sunlight. Glance up and wince at the glaring brightness.

Walk out of doors under a clear, blue afternoon sky. The luminous clarity almost overwhelms. The world around us appears remarkably crisp. Feel the gloom of early twilight in winter, how hard it is to relinquish sleep in the morning darkness.

Visit a museum and marvel at the artistry of painters capturing the magic of daylight through the centuries.

Daylight in its many natural renderings illuminates every part of our lives. This is the poetic aspect of daylight. The prosaic and practical is just as significant.

FIGURE 8.1 Sunlight filtering through a canopy of trees
Photo by Dodge + Burn Photography

FROM THE OUTSIDE IN

Imagine a home—or just the kitchen and bath—without windows, without daylight. It is uncomfortable at the very least . . . and illegal without some other type of effective ventilation.

Before the Industrial Revolution made glass widely affordable, windows were a luxury in cold climates. Even in warmer environments, the necessity of protecting openings from weather, insects, and intruders made large fenestration more expensive than the alternatives.

Our windows serve five important functions:

1. View
2. Comfort
3. Health
4. Beauty
5. Light

View

There are several ways to open a home to daylight: windows, clerestories (windows above eye level), and skylights.

FIGURE 8.2 The glass door and windows in this kitchen provide ample daylight and a view of the natural outdoor setting.
Design by Savena Doychinov, CKD, Design Studio International Kitchen & Bath, LLC, Falls Church, VA Bob Narod Photographer

Windows welcome in nature (see Figure 8.2), street life, and the neighborhood. To many, view, with the possibility of unexpected sights, the potential to discover threats, the relaxation offered by outside beauty, as well as the daylight they provide, is what windows are for.

Comfort

Windows also provide comfort. The open window lures in the gentle breeze that cleans a room of unwanted odors and provides the cooling passage of air over your skin. Windows also allow for the escape of tiresome heat during the summer months or can introduce a bit of fresh air during the winter months.

Health

Light—mostly daylight—affects our health in many ways. Notably, cyclical exposure to light entrains our circadian system, influencing our desire to sleep or wake, our alertness, even our hunger. Human biology responds to diurnal patterns of dark and light (earthly rotation). Inadequate light can lead to what is known variously as seasonal affective disorder (SAD) or as winter depression. Inadequate darkness may diminish the capability of our immune systems.

Most homes do not provide enough daylight to substitute for an outdoor environment. Nevertheless, effective design can support healthy circadian stimulation. And a well-designed home can thoroughly shield sleeping quarters from the problems of unwanted nighttime illumination.

Daylight, of course, also poses health risks, principally through exposure to ultraviolet radiation. This risk needs to be mitigated by limiting the time of exposure, especially to bright sunlight. Our homes provide this protection when we are indoors.

Beauty

The beauty of windows is not just what they provide to the interior but also how they articulate and decorate the exterior structure (see Figure 8.3). Without fenestration, most elevations look plain, almost unfinished. Little wonder that architects like to penetrate the exterior walls, sometimes at the expense of useful cabinetry on the interior.

Light

Windows, of course, admit daylight along with the view and perhaps the breeze (see Figure 8.4). Clerestories and skylights bring in light without the view. On a bright day, electric light may not be required at all (until darkness inevitably descends with Earth's rotation).

Is daylight the best light for indoor living? Judging from how often we open the blinds and turn off the electric lighting, the preference for daylight seems clear.

FIGURE 8.3 This home features several large windows adding to the beauty of the design.
Design by Habitat Studio, Edmonton, AB
Photo by Merle Prosofsky

FIGURE 8.4 The windows in this breakfast nook add to the enjoyment of the space by admitting daylight, a view, and fresh air.

Design by Jane Lockhart, Jane Lockhart Interior Design Toronto, ON
Photo by Brandon Barré

But remember how hard it is to read comfortably out of doors with bright beams of sunlight and deep shadows. Perhaps the best we can say is that daylight is very effective—as long as it is well controlled and available when we need it.

DAYLIGHT IN DIFFERENT FORMS

Recognizing all of the different experiences we have with daylight, we can see that it is, in fact, more than one type of light. Two basic qualities make up what we call daylight:

1. Direct beams of light from the sun
2. Diffuse reflection of sunlight through the sky

Sunlight

The direct rays of light from the sun provide a highly *directional* quality of light with high thermal content. Despite the *feel* of a beam of sunlight, the color of high-angle noon sunlight is relatively cool, about 5,500 Kelvins (K).

Skylight

When sunlight passes through molecules of air and moisture on a clear day, shorter wavelengths are scattered more than longer wavelengths. This is known as Rayleigh scattering and

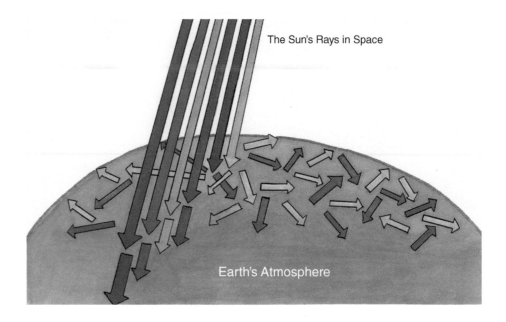

The Sun's Rays in Space

Earth's Atmosphere

FIGURE 8.5 Rayleigh scattering

produces the very cool, blue color of skylight as seen in Figure 8.5. Daylight received from a clear blue sky has a very high color temperature, which may exceed 10,000 K, depending on orientation to the sun.

Overcast Sky

When sunlight and skylight combine with cloud cover, daylight becomes quite diffuse, losing clarity, focus, and sparkle (see Figure 8.6). Although the character of overcast sky light differs from sunlight, color temperature remains relatively close, about 6500 K.

Reflected Daylight

Much of the daylight received indoors reflects off of the ground, which affects both the quantity and the color of the light. Gardens, patios, pools, decks, and pavement (see Figure 8.7) each reflect light differently.

CHARACTERISTICS OF DAYLIGHT

Daylight is dynamic—constantly changing. To better understand the various and varying aspects of daylight, we look four characteristics:

1. Intensity
2. Direction and texture
3. Color
4. Change

Intensity

Bright sunlight can deliver over 10,000 footcandles outdoors. In the shade, a clear sky at midday might deliver 2000 footcandles, a level 50 to 100 times what is delivered by typical indoor electric lighting.

Daylight reaching interiors through a window produces significantly lower illuminance than outdoors. Nevertheless, the daylighted area can be bright enough to cause glare and to overpower light in adjacent areas, making then seem dim by comparison.

FIGURE 8.6 Daylight becomes more diffuse on a cloudy day.
Photo by Dodge + Burn Photography

Direction and Texture

In the morning and afternoon, when the sun is relatively low in the sky, daylight arrives at an angle that easily penetrates windows, painting sharp shadows, creating harsh glare for those looking in that direction, and adding heat to the space.

In the afternoon, without direct sunlight and particularly with overcast conditions, daylight assumes a diffuse quality, with soft shadowing and more comfortable brightness.

Color

The color of daylight changes according to the proportion of direct sunlight, diffused skylight, and cloud cover. Thus, the color of daylight differs throughout the day. Geography, seasonality, and exterior reflection also affect the color.

Apart from the hour following sunrise or preceding sunset, when the color temperature of daylight is about 3000 K, daylight has a cool appearance, 5000 K or higher. Most residential electric lighting has a color temperature of 3000 K or lower.

The spectrum of daylight covers all visible wavelengths and extends well into the ultraviolet and infrared regions (hence the risk of sunburn, melanoma, and simple heat gain). Just as sunlight differs from skylight, so daylight presents different spectra. Daylight around noon

FIGURE 8.7 Light reflects indoors from the ground surfaces outside.
Design by Laurie Belinda Haefele, codesigner Colin Dusenbery, Haefele Design, Santa Monica, CA
Photo by Erhard Pfeiffer

offers a balanced spectrum; the spectrum of afternoon skylight is skewed significantly toward shorter-wavelength radiation.

Change

Some change in daylight is predictable: For example, the rotation and orbit of Earth around the sun manifests to our view as the daily and seasonal path of the sun. Other changes offer a degree of surprise; these include the arrival and departure of cloud cover and their myriad variations of pattern and density.

For many of us, it is the dynamic—changing the intensity, direction, texture, and color—that makes daylight so valuable.

ADMITTING DAYLIGHT

To use daylight, you need to bring it indoors . . . and control it. The more uniformly daylight is distributed around a space, the more comfortable and useful it will be. Conversely, the more daylight is concentrated, the more likely it will prove uncomfortably bright and warm.

Daylight comes with its own challenges, of course: chiefly, the glare and heat gain of direct sunlight. Further, large areas of glazing allow heat loss in winter. And every puncture in the exterior skin of the building increases the risk of water and moisture intrusion.

The amount and distribution of daylight indoors depend significantly on how a space orients to the sun, how it is admitted via windows, clerestories, and skylights, and how it is reflected within the space.

Glazing for Light and Heat

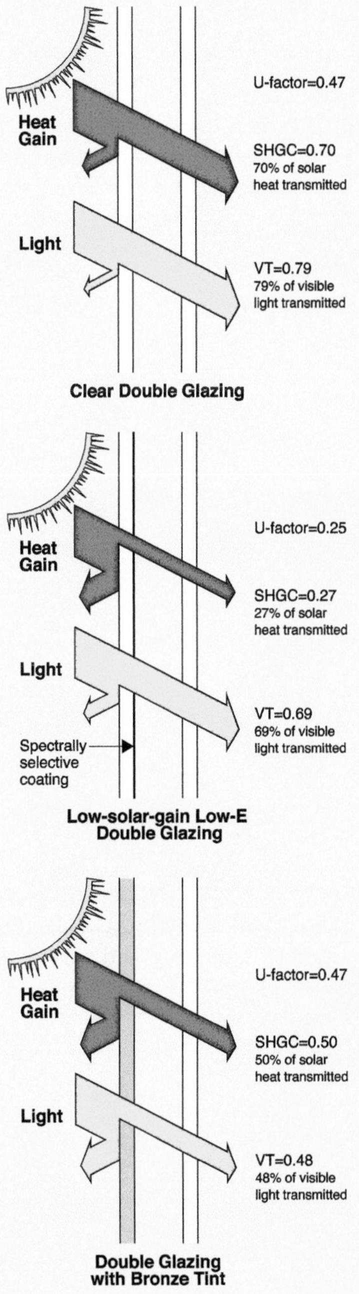

U-factor=0.47

SHGC=0.70
70% of solar
heat transmitted

VT=0.79
79% of visible
light transmitted

Heat Gain

Light

Clear Double Glazing

U-factor=0.25

SHGC=0.27
27% of solar
heat transmitted

VT=0.69
69% of visible
light transmitted

Heat Gain

Light

Spectrally selective coating

Low-solar-gain Low-E Double Glazing

U-factor=0.47

SHGC=0.50
50% of solar
heat transmitted

VT=0.48
48% of visible
light transmitted

Heat Gain

Light

Double Glazing with Bronze Tint

Many glazing materials offer a choice of light transmission, heat, and cost. Light transmission is measured by *visible light transmittance* (*VLT*) or visible transmittance (VT), the percentage of visible energy that passes through the glass.

Heat gain is generally measured by the *solar heat gain coefficient* (*SHGC*), which measures the portion of solar radiation that is transmitted either through direct transmission or through absorption and re-radiation.

Light–to–solar gain (LSG), the ratio of VT and SHGC, measures the effectiveness of glazing at providing light without heat. A high value for LSG would be good in warm climates but not in colder ones, where solar gain is valuable for winter heating.

Daylighting performance can be enhanced by a variety of techniques. These include use of low emissivity or low-e glass, tints applied to the glass, and other advanced window technologies that use an internal layer of insulating air between the glass panes. (See Figure 8.8.) Optimal design depends on geographic location, orientation, glazing area, and the value of the view.

FIGURE 8.8 VT, SHGC, and LSG value
Courtesy of Efficient Window Collaborative

Orientation

How a home and its rooms are sited determines the orientation to daylight. East- and west-facing spaces with windows enjoy direct sunlight throughout the year . . . along with the glare and heat gain that accompanies the light. In the northern hemisphere, south-facing spaces receive sunlight more in the winter, when the sun's trajectory is lower in the sky. North-facing spaces mostly have a sky view without direct sunlight.

In the southern hemisphere, of course, the northern exposure receives the sunlight. Near the equator, the north and south exposures experience more similar effects.

Seasonality means that a space receives more daylight during summer hours. A sunny summer kitchen is often dim in winter for breakfast and dinner preparation. Figure 8.9 illustrates the difference in the sun's trajectory during summer and winter.

All of this variation challenges daylight design and its integration with electric lighting.

Windows

Fenestration—windows and clerestories—admit light according to their size and height (for a given orientation and time of day). Larger openings let in more light. Taller openings allow light to penetrate deeper into a room, distributing light more effectively than shorter ones.

Windows—because they serve our desire for view— typically are oriented at eye level and proportioned to the ceiling height. Higher ceilings with taller windows provide more light and cover more room area than shorter windows that are common with lower ceiling heights.

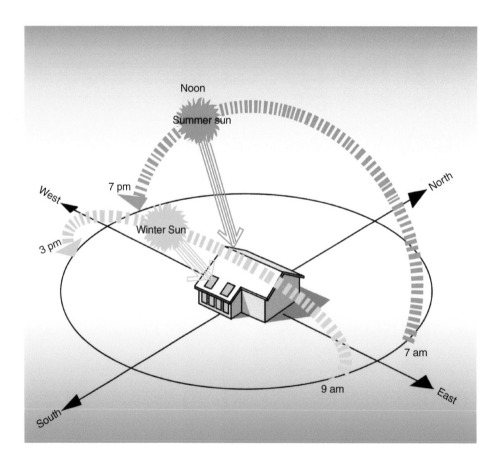

FIGURE 8.9 Sun trajectory

How Far Does Daylight Penetrate?

Areas near windows receive more light than those farther away. With a beam of sunlight, the effect is more pronounced than with a clear or overcast sky. Moreover, the *visual* effect is stronger than the measured effect. That is, the patch of daylight seen through the window as well as adjacent area illuminated by the window both make an interior wall or horizontal surface appear dim by comparison.

A common rule of thumb, illustrated in Figure 8.10, estimates effective daylight penetration at about 2.5 times the height of the window. The higher the window, the deeper the daylight will penetrate.

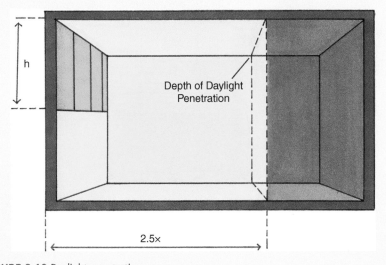

FIGURE 8.10 Daylight penetration

Clerestories

Clerestories, fenestration set above eye level, admit light but do not provide view (see Figure 8.11). Depending on the design of the building, they may also free up wall surface for cabinetry and other uses (see Figure 8.12). Higher walls and high ceiling heights are needed to use clerestories for daylighting.

Skylights

Skylights admit light through the roof. The quantity and quality of daylight they deliver depends on the design of the skylight itself.

Simple domed perforations in the roof admit both sunlight and skylight from all directions with little control over glare. (See Figure 8.13.)

The channel between the skylight dome and the opening into the space below is called a *well*. A deep well blocks direct view of the sun but also constricts the daylighted area of the room. The sides of the well reflect sunlight, acting as secondary light sources (see Figure 8.14). A well with splayed walls spreads daylight farther than one with vertical walls (see Figure 8.15).

FIGURE 8.11 The clerestory in this small bathroom brings daylight into the bath/shower enclosure while preserving privacy.

Design by Cheryl Kees Clendenon, In Detail Interiors, Pensacola, FL
Photo by Greg Riegler

FIGURE 8.12 The clerestories in this kitchen free up wall space for cabinets and the ventilation hood.

Design by Robin Rigby Fisher, CMKBD, CAPS, Robin Rigby Fisher Design, Portland, OR
Photo by PhotoDesign Inc.

FIGURE 8.13 Simple domed skylight
Courtesy of Velux

Skylights oriented away from the sun's path (north facing in the northern hemisphere) are called *monitors* (see Figure 8.16). The typical monitor admits skylight through a vertical aperture (as opposed to a skylight's horizontal aperture) and reflects the light into the space.

Because the monitor does not see direct sunlight, the daylight admitted is more comfortable and significantly cooler in color than that admitted from a horizontal skylight.

Solar tubes (see Figure 8.17) are another option used to bring natural light into homes. These cylinders are commonly used in bathrooms and some kitchens and can even be used on the first floor of a two-story house.

FIGURE 8.14 Skylight
Courtesy of Velux

Sharing Daylight

Windows, clerestories, and skylights all localize daylight (as will any luminaire). To illuminate a space more fully, the daylight needs to be shared, or spread around. Three techniques help accomplish this goal:

1. Reflective finishes
2. Reflecting surfaces
3. Interior glazing

Reflective Finishes

Reflective finishes help to transfer daylight from its source—a window, for example—deeper into the space. Light colors on the ceiling play a particularly important role. With a high-reflectance white ceiling (r = 90 percent), more than half of the daylight illumination away from the window is reflected rather than direct. High reflectances brighten interior surfaces, creating secondary sources of daylight that soften shadows (see Figure 8.18).

Reflecting Surfaces

How does daylight reach the ceiling (since daylight originates well above the ceiling plane)? As Figure 8.19 shows, daylight first reflects off of the exterior ground, deck, patio, or garden and then it reaches interior ceiling.

FIGURE 8.15 Skylight with splayed well
Courtesy of Velux

Daylight also can reflect from interior horizontal surfaces designed to transfer it into the space. A typical approach is a light shelf that intercepts some of the incoming daylight and redirects it up and into the space (Figure 8.20). Light shelves combine with tall windows or clerestories to make an especially comfortable and effective system for daylight control and distribution. As Figure 8.21 illustrates, the light shelf also blocks some direct sunlight, shading the hottest area near the window.

FIGURE 8.16 Monitor skylights
Courtesy of Velux

Interior Glazing

Interior spaces can borrow daylight (and electric light) if the partitions are glazed with transparent or translucent material rather than enclosed with wallboard or other opaque material.

For example, a bathroom with a window may be adequately lighted during the day without using its electric lighting. If a shower is enclosed by glass (see Figure 8.22) in a bathroom with a window, it may not need any supplemental lighting during the day.

Shading

Direct sunlight can impose a significant heat and glare burden on interior spaces, a burden that needs to be addressed.

Architectural design can minimize these problems by siting the building, and its primary fenestration, to favor skylight over direct sunlight or by adding permanent sun screens or screens around windows (see Figure 8.23). In some geographical areas, these architectural considerations can assume paramount importance.

Effective modeling of architectural design can be done with computer programs (simple or sophisticated) that plot sun trajectories at different times of the year. Even with architectural shading, adjustable window-based shades may be required for adequate daylight control.

FIGURE 8.17 Solar tube
Courtesy of Velux

In more temperate areas, issues of view and aesthetics may take precedence. Here, adjustable shades alone may be a more appropriate approach (see Figure 8.24). Not every window needs to be shaded, but every window needs to be considered.

INTEGRATING DAYLIGHT AND ELECTRIC LIGHTING

Daylight, with its many benefits, is irreplaceable. And, while Earth rotates and orbits with consistency, the daylight it delivers varies all the time. Sometimes the light is ample; sometimes it is not.

FIGURE 8.18 A white ceiling provides reflectance in this kitchen where the source of daylight is limited to a single window.
Design by Germán Brun, LEED AP, codesigner Lizmarie Esparza, Den Architecture, Miami, FL
Photo by Greg Clark

Do you envision a kitchen illuminated entirely by electric lighting as it might be on a late afternoon in winter or a spring evening? Or do you imagine the kitchen lighted by a flood of daylight? Of course, as the day and seasons pass, the balance between daylight and electric light changes.

The ideas that follow suggest starting with the daylighted environment, identifying where that light is insufficient or perhaps overly bright, and then combining daylight and electric light to address these issues.

Compensating for Insufficient Daylight

Imagine a kitchen in the middle of a clear afternoon. What parts of the room are inadequately lighted despite the abundance of daylight? Perhaps it is counters that are shadowed by overhead cabinets. Maybe it is a vanity in a far corner of the bathroom. Now imagine the space when daylight is limited by the time of day, the winter clock, or cloud cover. What other areas now feel dim? Then it is dark, perhaps in the late afternoon, perhaps well into the evening; without daylight, only electric lighting provides illumination.

FIGURE 8.19 Indirect daylight

FIGURE 8.20 Window with light shelf

FIGURE 8.21 Clerestory with light shelf

FIGURE 8.22 A transparent shower partition allows daylight and electric light from the main bathroom space to light the shower space

Design by Laura Stein, codesigner Natalie Graziano, Laura Stein Interiors, Inc., Toronto, ON Photo by David Bagosy Photography

FIGURE 8.23 Sun screen

FIGURE 8.24 Adjustable shades

TABLE 8.1 Daylight and Electric Light

Characteristic	Daylight	Electric Light
Intensity	Bright to dim	Moderate to dim
Direction	From windows and skylights	From ceiling, cabinet, pendant
Color	Generally cool, full spectrum	Generally warm, limited spectrum
Change	Independently	By command

Although the entire space is can be illuminated by electricity, not all of the electric lighting is needed all of the time. Distinguishing which layers of lighting are needed when allows you to control each layer separately and operate it only when it is needed.

Overcoming Bright Daylight

As we discussed earlier, bright daylight near windows can make interior walls appear dim by comparison. In addition to shading against glare and heat gain, you may need to add electric lighting at the interior of a daylighted space to balance brightness around the room.

This is another layer of lighting that benefits from separate controls so that it can harmonize with different daylight and electric lighting conditions.

Combining the Effects

Daylight and electric light behave differently in terms of their key characteristics: intensity, direction, color, and change (see Table 8.1). Combining their effects takes some nuance. We sketch out the issues here and address them again when we conceptualize and then develop a lighting design in Chapter 15, "Documenting the Lighting Design."

When daylight is ample, spaces can be quite bright, and that brightness is difficult to replicate with electric lighting. "Difficult" here means costly in terms of energy use and complicated in terms of luminaire design and location.

Because a skylight enclosure reflects like a mirror at night, it is extremely difficult to light up a skylight well to match how it appears during the day.

The color of daylight changes throughout the day and the year. Except with special color-changing equipment and dimmed incandescent lighting, electric lighting looks the same.

These challenges suggest that any kitchen or bath with windows or skylights should be envisioned under daylight (and supplemental electric lighting) and also under electric lighting.

Controlling Electric Light

By now it should be clear that careful control of electric lighting is required when integrating electric lighting and daylighting so that light is available when and where it is needed.

The most important step is dividing the electric lighting into separate controllable zones so that specific areas can be illuminated as needed (rather than activating all electric lighting in the room when daylight dims a little). Variable control (dimmers) is more flexible than simple on/off control (switches). We look at both of these techniques in depth in Chapter 13, "Lighting Controls."

Looking for Daylight

Use your lighting journal to record your observations in this exercise. You can record your outdoor experiences in any location with adequate daylight. For the interior experiences, you can use your own kitchen or bath, other rooms in your residence, or your workplace.

1. For each of these exercises, note your location and surroundings, then describe the quality of daylight you experience in terms of intensity, direction (think shadows), and color.
   a. Outdoors in the morning
   b. Outdoors at noon
   c. Outdoors in the early evening, before dusk
2. Review your notes from outdoors at different times of day. What do you conclude from your observations?
3. Select and describe an interior location that receives daylight at different times of the day. Identify the surfaces that are well lighted and those that are dim or in shadow; describe how daylight influences the color and texture of key materials in the space; and note whether you experience glare or discomfort when:
   a. The location is brightly lighted.
   b. There is less daylight.
4. Compare your experiences with daylight of high and low brightness. What differences do you note? What similarities?

SUMMARY

Daylight has broad influence over human experience beyond simply illuminating interior spaces. Issues of view, comfort, health, and beauty affect how daylight is designed into homes.

Daylight takes several forms, depending on the proportion of sunlight and skylight and the degree of cloud cover. The intensity, direction, texture, and color of daylight change throughout the day and year. Except near sunrise and sunset, daylight is generally stronger and cooler in color than most electric lighting.

Windows, clerestories, and skylights all admit daylight and do so to different effect. Distribution of light around a space and control over glare and heat gain are important for successful daylighting. Because daylight is not likely to meet all lighting needs all of the time, electric lighting needs to be integrated into building plans. Controlling layers of electric light separately provides better integration than simply turning the room lights on and off.

REVIEW QUESTIONS

1. Discuss the different aspects of daylight in the human experience. (See "Daylight and Human Experience" page 81)
2. What are the different forms of daylight? How does sunlight differ from skylight? (See "Daylight in Different Forms" page 85)

3. What are the basic characteristics of daylight? How does daylight compare in general to electric light? (See "Characteristics of Daylight" page 86)

4. Describe the different methods of admitting daylight into the interior. Why are high-reflectance surfaces important? (See "Admitting Daylight" page 88)

5. Why might we need electric lighting during the day? What are some challenges in integrating daylight and electric light? (See "Integrating Daylight and Electric Lighting" page 97)

Schematic Design

Schematic design for lighting establishes the basic concept in terms of the activities to be lighted, how the space will feel when it is lighted, and how the lighting integrates with all of the other design elements. The schematic design provides the foundation and organization for the details of light source, fixture, and control, and the specific locations of each.

Learning Objective 1: Program lighting for kitchen and bath applications.

Learning Objective 2: Develop lighting concepts based on activities and experiences.

Learning Objective 3: Develop lighting concepts that integrate with architecture.

Learning Objective 4: Communicate lighting concepts.

PROGRAM

Lighting—like all design—begins with the program: project scope, the client's functional and aesthetic objectives, the constraints that limit choices, planned activities and tasks, and the preferences and priorities that can significantly impact the budget.

A clear definition of the program helps to define the design problem and initiate a schematic approach to the lighting.

Scope

New construction allows the most design freedom; renovation projects can impose few or many limitations, depending on how deep into the building structure the renovation extends.

Opening the ceiling and walls in a renovation provides more freedom for lighting placement but you never know what you are going to find during a renovation. Lighting costs may go up or down depending on the conditions.

Clients

Age and visual capability are key client attributes in terms of lighting. Older and less visually capable users need more light for visual tasks and better control over glare than younger ones. Families with children may have different needs than singles or couples without children in terms of simplicity, clean up, even access to first aid information.

Activities and Tasks

Kitchens and baths host numerous tasks. For kitchens, an important consideration is the how often the space will be used for activities beyond food preparation—breakfast, conversation before dinner, household accounts, and so forth (see Figure 9.1). For bathrooms, the tasks and the users typically vary, depending on whether the space is intended as a master bath, children's bath, or guest bath. The lighting approach can differ markedly from one to another.

Preferences and Budget

Aesthetic preferences and budget clearly influence the lighting program, as they do every other aspect of a project. Both will affect decisions about whether and how to integrate lighting with the architecture as well as specific luminaire selection.

SCHEMATIC DESIGN

Schematic lighting design establishes a concept for lighting a kitchen or bath. As a concept, the details have yet to be fully developed. Nevertheless, you should be able to *describe the lighting effects and construction implications* of the scheme . . . and secure client approval to develop the idea so that it can be built.

FIGURE 9.1 This large open kitchen designed with multiple work and seating areas is an inviting space for social interaction and activities.
Design by Claire Reimann, AKBD, Jason Good Custom Cabinets, Victoria, BC
Photo by Joshua Lawrence

The schematic design should answer three questions:

1. How are the key *activities* in the space illuminated?
2. What visual *experiences* will users of the space enjoy?
3. Where (and how) is lighting integrated into the *architecture*?

To create the scheme, we use the concepts and vocabulary learned in the preceding chapters, beginning with a general discussion, then moving to the different specifics for kitchens and bathrooms.

Schematic design is a thought process, not a recipe. Solutions arise from the specifics of each space and client. Compromises are inevitable.

Layers of Light

Since kitchens and bathrooms are task-oriented spaces, we start by considering a lighting layer oriented to local activities, such as food preparation or grooming; this is known as task lighting. Then we consider what additional lighting is needed to help orient people to the space and facilitate movement through it; this is called ambient lighting. Finally, we will look at how to make the space feel pleasant and attractive; this is focal and decorative lighting.

This sequence suggests client preference for functionality (e.g., a cook-in kitchen or a social kitchen); more than for aesthetic impact (e.g., a master bath [see Figure 9.2] or a guest bath; [see Figure 9.3]).

FIGURE 9.2 Master bath

Design by Shawn McCune, CKD, Kitchen Design Gallery Inc., Lexexa, KS

Photo by Bob Greenspan

FIGURE 9.3 Guest bath
Design by Christine Pandur, Design Eye Ltd., Edmonton, AB

The Lighting Concept

Once you and the client agree on the lighting concept, you develop the design so that it meets the program objectives within the project constraints and document it so that it can be built. But how do you agree on the concept in the first place?

Communicating the concept is the critical last step in schematic design (see Figure 9.4). To finish this chapter, we explore some techniques for *showing light* effectively (see "Communicating the Concept").

LIGHTING FOR KITCHEN ACTIVITIES

The lighting concept needs to consider the various visual tasks, their specific locations, and how they should be illuminated. For example, a kitchen peninsula may support both food preparation, which takes place on the horizontal surface and can require as much as 50 footcandles of measured light, or social activity, which takes place across the peninsula and needs gentler vertical illumination that softens shadows on people's faces (see Figure 9.5). You may find it helpful to review Chapter 4, "Seeing the Work."

Food preparation, dining, paperwork, conversation, and socializing represent some of the principal activities that take place in the kitchen. In addition, food and household products move into and out of the kitchen, surfaces are cleaned, and people simply walk around.

PERSPECTIVE-OPTION #1

FIGURE 9.4 A rendering helps to communicate the concept to the client.
Design by Kim Van Ruskenveld, AKBD, Design Eye Ltd., Edmonton, AB

Food Preparation

The key food preparation activities in the kitchen typically take place on horizontal surfaces: countertops, sinks, cooktop, island, or table (where applicable) (see Figure 9.6). These spaces are usually (1) located directly under a cabinet, (2) covered by a range hood, or (3) open to the ceiling above.

Conceptually, this suggests three approaches:

1. For work surfaces directly under cabinets, locate lighting on the underside of the cabinet (under-cabinet luminaires). This approach has several advantages. Light travels only a short distance to the surface so it concentrates on the task, avoiding wasted spill (and energy). Luminaires set below eye level (for most standing adults) do not create direct glare. Luminaires directly above the task and in front of the person at work do not create body shadows.

2. For cooktops with range hoods, use luminaires built in to the hood (see Figure 9.7). Since this approach, while practical, may prove insufficient, supplemental, ceiling-based lighting may be needed as well.

FIGURE 9.5 Kitchen with pendant lighting at the peninsula
Design by Diane Foreman, CKD, CBD, codesigner Jamie Rupprecht, AKBD, Neil Kelly Design Build/Remodeling, Lake Oswego, OR
Photo by Brinkman Photography

3. For work surfaces open to the ceiling, lighting either emanates directly from the ceiling (e.g., from pendant or recessed downlights) or reflects off the ceiling indirectly (see Figure 9.8). Direct lighting concentrates light on the work surface but may create shadows (and glare). Indirect lighting is generally more comfortable and free of shadows. But, as it spreads light throughout the room, indirect lighting may require more energy to provide sufficient illumination at the work surface.

Do not overlook tasks at the sink, which require ample illumination. Sticky food particles on pots, pans, and dishes are often difficult to see due to their small size and low contrast, and the work typically is done in haste to be finished. A single luminaire over the sink is an obvious approach, but shadows may be an issue. Spreading the light out may provide a more comfortable solution.

Socializing

Effective lighting for social activities favors people's faces and connects the participants in conversation or shared activity. For social activity, the quantity of illumination is much less significant than its quality.

Soft, diffused light, especially from in front of faces, flatters people and helps us to read facial expressions (see Figure 9.9). Concentrated light from overhead typically casts disturbing shadows over faces.

FIGURE 9.6 A variety of light sources illuminates the food preparation surfaces in this kitchen.
Design by Peter Ross Salerno, CMKBD, codesigner Diane Durocher, Peter Salerno Inc., Wyckoff, NJ
Photo by Peter Rymwid Architectural Photography

Arranging appropriate lighting is easier where social areas are separated from food preparation (e.g., a breakfast table [see Figure 9.10] or breakfast nook [see Figure 9.11]) than for spaces that share social and food preparation tasks (e.g., a peninsula).

To accommodate social activity from a breakfast coffee to a nightcap, you need dimming control, including the ability to adjust various layers of lighting separately.

Storage

While food preparation and socializing represent the two crucial kitchen activities, a lot more goes on. People load and unload storage areas and the dishwasher, clean the floor (task surface cleanup is part of food preparation, of course), and roam at night to raid the refrigerator.

Most of these activities can be lighted effectively using ambient, ceiling-based approaches. With the flexibility of dimming control, a single layer of illumination can be both bright enough for cleanup and sufficiently dim for a nighttime trip into the kitchen.

The top shelves of cabinets or pantry storage can present challenges, however. For these applications, it is important to use lighting with a wide distribution or to locate lighting equipment close enough to the shelves so that light reaches into the storage areas (see Figure 9.12).

FIGURE 9.7 Range hood with built-in lighting.

Design by Tammy MacKay, AKBD, Design Eye Ltd., Edmonton, AB

FIGURE 9.8 Light reflects off the ceiling in this kitchen featuring light surfaces and ample daylight.
Design by Tina Lynne Muller, Drury Design Kitchen & Bath Studio, Glen Ellyn, IL
Photo by Eric Hausman

LIGHTING FOR BATHROOM ACTIVITIES

Principal activities in the bathroom include grooming, showers and baths, cleaning the space itself, and all the rest. In some bathrooms, these activities are compartmentalized; in others, they are simply separated by the plumbing that serves them.

Importantly, as the lighting requirements differ, so will the conceptual solutions.

Grooming

Due to the small size and (often) low contrast of the task, the illumination levels required for grooming are among the highest typically found in homes. Light sources should be diffuse rather than concentrated. This suggests fluorescent technology, which delivers diffuse lighting with minimal heat contribution and can provide satisfactory color rendering if chosen carefully.

Facial grooming in front of a mirror—whether for makeup or shaving—is the quintessential vertical task. For best visibility, lighting should approach the face from in *front*, rather than from above, where direct lighting creates harsh shadows.

FIGURE 9.9 The chandelier in this kitchen provides soft, diffused lighting over the table where meals and socializing take place.

Design by Lauren Levant Bland, Jennifer Gilmer Kitchen & Bath Ltd., Chevy Chase, MD
Photo by Bob Narod, Photographer, LLC

There are several approaches to illuminating a mirror:

- Light directly from around the mirror (see Figure 9.13).
- Light via luminaires on both sides of the mirror. Placing luminaires on both sides of the mirror provides the most effective illumination (see Figure 9.14).
- Light from above the mirror. Light above the mirror, even when elongated, can create unpleasant shadows.
- Light indirectly from around the mirror. Concealing luminaires behind the mirror surface provides more diffuse and comfortable illumination than exposed luminaires (see Figure 9.15). Construction is a little more complicated, and indirect light does require more initial lumen creation to deliver adequate amounts onto the face.
- Light indirectly from a cove over the mirror. Indirect illumination is very comfortable but may lack focus. Since it reflects off the ceiling overhead, this approach still creates some shadows.
- Light directly from a soffit over the mirror, using a large diffuser. In addition to some shadowing from the overhead source, this solution may remind clients of a hotel stay.
- Light from the ceiling overhead. Besides the shadows, locating lighting in the ceiling tends to either backlight—or silhouette—the person at the mirror or to reflect in the mirror itself, depending on where the luminaires are placed (see Figure 9.16).

Shower

For some people, light "borrowed" from the rest of a brightly illuminated bathroom is sufficient for a small shower with a transparent glass enclosure or curtain. In this concept, no additional lighting is needed.

FIGURE 9.10 A pendant fixture provides the lighting for this dining area.
Courtesy of Kichler

FIGURE 9.11 The separation of this breakfast nook from the food preparation area expands the choice of lighting design options.

Design by Tammy MacKay, AKBD, and Christine Pandur, Design Eye Ltd., Edmonton, AB

FIGURE 9.12 The placement of the recessed lighting in this kitchen design provides light for the cabinet interior as well as the countertops.
Design by Beverley Leigh Binns, Binns Kitchen + Bath Design, Pickering, ON
Photo by Tim McClean

Other people prefer to have a more focused light, especially for shaving while showering. Illumination can be provided simply by a central downlight (typically, one designed for this purpose, such as an enclosed and gasketed type) or, more elegantly and elaborately, from luminaires embedded in the walls.

Baths and Spas

Showers today no longer suggest a utilitarian lighting concept. As with baths and spas, luxury and a luxurious lighting concept can be incorporated into the shower design (see Figure 9.17). The more lavish the environment, the less mundane the lighting.

Softly colored light concealed in a cove, perhaps with a slow fade to change color, can create a very special sensibility for the bath.

Accent lighting, dimmed to an intimate level, is another approach to make time in the bath more pleasant (see Figure 9.18).

Remember, overhead lighting will likely be in direct view of the person lying in the tub and therefore should be comfortable to the eye and controlled by a dimmer.

FIGURE 9.13 Lighting at the mirror is provided by an overhead recessed light as well as pendants on both sides suspended at different levels.

Design by Sol Quintana Wagoner, codesigners John Kavan and Vanessa Tejera, Jackson Design and Remodeling, San Diego, CA
Photo by PreviewFirst.com

FIGURE 9.14 Lighting on both sides of the mirror provides effective illumination.
Design by Thomas Richard Kelly, CMKBD, Marblehead, MA, Nokomis, FL, Belgrade Lakes, ME

Cleanup

Like the kitchen, the bath is a room that needs to be kept clean and hygienic. Since many bathroom areas are hard to see and reach (think of the floor behind the pedestal of the sink or toilet), it is important to provide adequate lighting for cleanup.

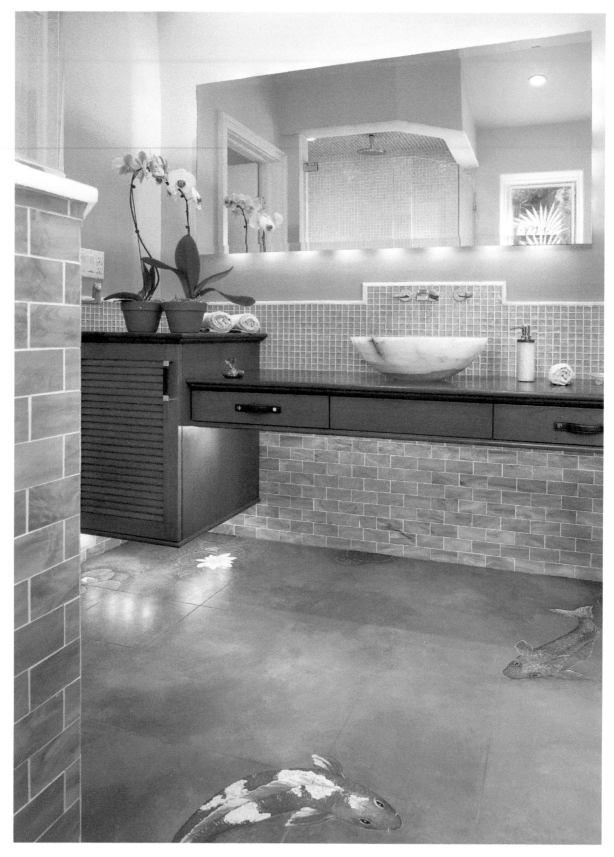

FIGURE 9.15 Lighting installed behind the mirror in this bathroom provides comfortable illumination.
Design by Tess Giuliani, CKD, Tess Giuliani Designs, Inc., Ridgewood, NJ
Photo by Peter Rymwid Architectural Photography.

FIGURE 9.16 In addition to recessed lighting in the ceiling, this bathroom design includes built-in lighting at both sides of the mirrors for grooming.
Design by Elizabeth A. Rosensteel, Elizabeth A. Rosensteel Design/Studio, Phoenix, AZ
Photo by Robert Reck

Shadows from plumbing fixtures can present notable challenges, particularly where lighting is provided by a single luminaire. Highly reflective surfaces help distribute light and mitigate shadowing.

First Aid

In most homes, the bathroom holds medicine and serves as the first aid station. At least one location needs plenty of light to see into cut fingers or to read the small print on medicine bottles (see Figure 9.19).

And the Rest

Using the toilet is not a visually challenging task . . . except that light for reading needs to be comfortable (not a harsh beam overhead) and lighting at night needs to be low enough to limit disturbance to sleep rhythms.

FIGURE 9.17 Colored light in shower

Design by Young Brothers Inc., Oklahoma City, OK
Photo by Keffer-Sharpe Photography

FIGURE 9.18 Accent lighting in the alcove in this luxurious bathroom can be dimmed to enhance relaxation during an evening bath.
Design by Paul Knutson, Knutson Residential Design, LLC, Saint Paul, MN
Photo by Troy Thies

A compartmentalized toilet needs its own lighting. An open plan may not, depending on the overall size of the room and the overall ambient lighting (see Figure 9.20).

EXPERIENCE

In terms of schematic design, the visual experience asks: How does the lighting make people feel about the room?

The visual experience of the space may be *spatial*: How does lighting reveal the form, finish, and detail of the space? Another aspect of experience is *emotional*: Should lighting impart a sense of intimacy, expansiveness, or efficiency? And finally: How does the lighting support the *aesthetic* aspects of the design?

Spatial Experience

While lighting affects how we experience any space, some rooms simply call out for a spatially focused lighting concept. Consider an open plan kitchen with a sloped, wood-planked ceiling, as seen in Figure 9.21. Ambient indirect lighting on the ceiling emphasizes the sense of volume and dramatizes the materials. In a galley kitchen that needs to feel larger than its limited footprint, lighting that brightens vertical surfaces conveys a sense of spaciousness.

As a designer, ask yourself if you prefer that a large room be held together by an ambient lighting system or separated into small spaces with more localized pools of brightness.

FIGURE 9.19 The pendant lights hung in front of the mirrors in this universal design bathroom provide ample light for grooming and reading the labels on medicine bottles.

Design by Shawn McCune, CKD, Kitchen Design Gallery Inc., Lenexa, KS
Photo by Bob Greenspan

Emotional Experience

Imagine a home with a master bath, children's bath, and guest bath. Should they all feel the same?

Many clients want the experience of comfort and luxury in the master bath. Lighting that supports such an experience will soften the brightness of the light sources and distribute them around the space, supplementing task light with touches of decorative illumination (see Figure 9.22).

The children's bathroom, in contrast, might emphasize cleanliness and efficiency—up and out of the house and do not leave it a mess. Here lighting might follow a more functional

FIGURE 9.20 A recessed light provides lighting for the toilet area. The opaque screen affords the user some privacy without blocking the light from the rest of the space.

Leslie Lamarre, CKD, CID, codesigners Erika Shjeflo and Casey Darcy, TRG Architects, Burlingame, CA Photo by Bernard Andre Photography

approach; perhaps all built in or, at most, simple luminaires. Of course, this concept might also appeal to an all-business client as well.

And the guest bath . . . perhaps utility is unimportant compared to the visual impact of sumptuous fixtures—both plumbing and lighting.

Aesthetic Experience

There are two chief aesthetic concerns in the visual experience:

1. What is the appearance of the lighting equipment itself? How should the light fixtures coordinate with other equipment? This is largely a matter of determining the appropriate style, scale, and detail desired . . . and then finding it among the myriad of fixtures available.

FIGURE 9.21 Lights installed along the wood beams help to illuminate this kitchen featuring a wood-planked sloped ceiling.

Design by Lauren Levant Bland, Jennifer Gilmer Kitchen & Bath Ltd., Chevy Chase, MD
Photo by Bob Narod, Photographer, LLC

2. How does the lighting reveal the architectural and decorative elements in the space? Is there art that needs to be illuminated? Are there details in the cabinetry that should be highlighted—perhaps special objects inside that deserve emphasis or an aspect of the millwork itself (see Figure 9.23), such as an illuminated rail or toe-kick?

FIGURE 9.22 The light reflecting off of the iridescent tile in the alcove provides decorative illumination in this bathroom.

Design by Janice Stone Thomas, ASID, CKD, codesigner Alia Richards, Stone Wood Design, Inc., Sacramento, CA
Photo by Dave Adams Photography

ARCHITECTURAL INTEGRATION

Before light reaches a task surface, human face, or architectural detail, it originates in a luminaire connected to the architecture of the space (or passes through a window or skylight).

FIGURE 9.23 Lighting in the decorative alcove above the cabinets draws attention to the plate on display.
Design by Angela Victoria Rasmussen, House 2 Home Design & Build, San Jose, CA
Photo by Dean J. Birinyi Photography

Sometimes it is easy to integrate the equipment so that it delivers the appropriate quality and quantity of illumination. Cabinet-mounted task lighting installed immediately above work surfaces offers a good example. Providing ambient illumination from higher or sloped ceilings can be more challenging (see Figure 9.24).

Daylight Integration

Conceptually, daylight integration divides into two questions: How should you light when daylight is present? and How should you light when daylight is not present?

When daylight is present, some surfaces and areas will be amply illuminated; others will not, or will not appear comfortably bright in contrast to much more brightly daylighted surfaces (see Figure 9.25). Interior vertical surfaces outside of the daylighted area can be especially problematic. And, of course, these relationships change over time.

With the daylighted and nondaylighted areas identified, electric lighting should be *divided and controlled* so that supplemental illumination can be provided as needed (and turned off, or dimmed, when not needed). Bright windows need to be shielded against the glare of the sun at low angles or the reflections off exterior materials.

Generally, electric lighting that supplements or replaces daylight will not match its intensity or its color and distribution around the space. The electric lighting will need to be comfortable and attractive in its own right—both when used in conjunction with daylight and when used alone (e.g., at night).

Exposed electric light sources tend to look warm, even yellow, compared to the spectrum of daylight. If this is unappealing, concealed sources may be a better approach for supplemental illumination.

FIGURE 9.24 Note the special lighting installation in this kitchen featuring a high, curved, wooden ceiling.
Design by Wendy F. Johnson, CKD, CBD, Designs for Living, Manchester Village, VT
Photo by Dennis Martin

Ceiling Conditions

Ceiling construction can influence the conceptual approach to ambient lighting significantly. Lighting can be recessed into the ceiling only if there is adequate plenum space above.

Downlighting in this way can provide inconspicuous, uniform illumination that focuses attention on horizontal surfaces. While the use of recessed lighting is a decorative decision with implications for the overall style of the space, the lighting equipment itself presents only limited aesthetic complications (size and finish).

Without depth for recessing, fixtures that provide ambient lighting must be exposed (either surface or pendant mounted) or installed around the perimeter in a cove or on top of cabinets. Pendant luminaires can deliver uplight, downlight, or a combination. Importantly, they represent a highly visible decorative element, one that requires consideration of the rest of the materials and finishes in the space.

Ceilings lower than 10 feet (or about 3 meters) can be awkward for pendant lighting unless it is suspended over a peninsula or island (see Figure 9.27). Sloped ceilings present their own challenges, as recessed lighting equipment may be difficult to reach and not adequately shield bright light sources from view. Suspended or cove-mounted lighting may offer a better approach.

Coves

If the concept for ambient lighting does not place equipment in, on, or suspended from the ceiling, the most practical alternative is integrating it into a cove or similar structure. This approach can also be used in addition to other ambient lighting equipment present in the design.

FIGURE 9.25 This kitchen is nicely illuminated with daylight. Note the recessed light to the right of the stove, which provides needed task lighting since the refrigerator blocks some daylight to the counter.

Design by Chris Novak Berry, codesigner Julie Gragg, Brooksberry Kitchens and Baths, St. Louis, MO
Photo by Alise O'Brien Photography

Ceiling Construction and Recessed Lighting

In a wood-framed home, ceiling construction usually features joists or trusses that support the floor above and ceiling below. The space between the joists and trusses provides room for recessed lighting equipment. (see Figure 9.26). The height of the joist, which depends on the distance that the joist spans and the weight it carries, determines the depth and width of the opening and limits the depth of the equipment that can fit in it.

FIGURE 9.26 Ceiling construction

Solid wood joists commonly range from 6 to 12 inches nominal (152–305 mm). Engineered joists, which can be spaced farther apart, tend to be a little taller. Manufactured floor trusses range in height from 12 to 24 inches (305–610 mm).

The deeper the opening, the greater the choice of compatible equipment. Shallow openings may require more costly equipment to provide the desired comfort and aesthetic appeal.

Wallboard ceilings attached to concrete structures (as in high-rise construction) may provide less than 4 inches (100 mm) of space and require ultra-shallow recessed or semirecessed luminaires.

Ceilings at the exterior of the home typically include thermal insulation to reduce heat and cooling losses. Such insulation can trap heat in a luminaire, so the fixture must be specially designed for this application.

FIGURE 9.27 Small-scale pendant lights were selected for the peninsula in this small apartment kitchen.
Design by Tim Scott, codesigner Carmen Mueller, XTC Design Incorporated, Toronto, ON

Lighting concealed in a cove or above cabinetry, as seen in Figure 9.28, provides uplight that makes the ceiling the brightest surface in the space and delivers soft ambient lighting without sharp (or any) shadows. As with recessed downlights, there is little challenge in coordinating the details of the lighting equipment with other elements in the design.

To be effective, a cove needs to be large enough for the light to escape from it. A cove that is too cramped will trap light and create hot spots on the wall and ceiling immediately above it. The smaller the lighting equipment, the smaller the cove can be and still distribute light effectively. The cove size can vary from the light source to the ceiling and to the wall behind, depending on the size of the light source itself.

FIGURE 9.28 Lighting above the cabinetry provides uplight for the ceiling.
Design by Wendi Zampini, Home Systems, Lafayette, CA

Where cabinets do not extend to the ceiling, a cove can be easily built on top by adding a short fascia at the front of the cabinet if you need to disguise the fixture. Coves can also be built behind or above bathroom mirrors, where they blend into the structure, or around the perimeter of the space (where they will be more prominent).

COMMUNICATING THE CONCEPT

From the program and an understanding of the activities, experiences, and architectural issues, a lighting concept emerges. It is not a recipe or a prescribed route; it is one or several of many possibilities.

To communicate the schematic design, you can describe it in terms of the key activities to be lighted, the feeling you intend to convey, and how the lighting integrates into the architecture. The layers of lighting and how they integrate with daylight and are controlled will help explain the concept.

Sketches showing how the lighting plays on horizontal and vertical surfaces generally work better than a detailed plan of a lighting layout (which has not been discussed yet). Of course, you need to have an idea of the location and type of lighting intended, but you do not need to be exact at this point.

You can create these images by coloring renderings of the space with a yellow marker to show where you expect lighting to go. Perspective views are particularly useful to begin to understand how the light will fall on vertical surfaces (see Figure 9.29).

Remember, in terms of visual experience, understanding where light goes is at least as important as where it originates!

FIGURE 9.29 Rendering showing lighting

Design by Kim Van Ruskenveld, AKBD, Design Eye Ltd., Edmonton, AB

Lighting Concepts

This exercise involves your own kitchen or another accessible space to create a new lighting concept. Imagine that you are doing a general renovation so that existing construction detail is not an issue. Use your journal to describe your concept in words; use sketches as appropriate to convey the concept.

1. Create the program by describing the scope of work, client, key activities, and preferences.
2. Express the program as a problem. For example, "The challenge of lighting this kitchen is _____ while minimizing/maximizing/ satisfying _____ [a constraint in the space or project].

3. Develop a concept for a layered lighting design including task lighting, ambient lighting, and lighting for emotional and aesthetic experience.
4. Explain how you will integrate daylight and electric light, indicating how the electric lighting is divided and controlled.
5. Explain how you will integrate each layer into the architecture of the space (ceiling, cabinetry etc.).

SUMMARY

Schematic design begins with the program—what needs to be accomplished. To move from the program to a design concept, you can use the layers of lighting to address the different activities, create visual experiences, and integrate lighting into the architecture.

In kitchens, understanding both functional and social uses helps to develop the lighting concept. In bathrooms, recognizing the different priorities in master baths, children's baths, and guest baths tends to produce more appropriate designs than a single treatment.

REVIEW QUESTIONS

1. What are the key work surfaces to light in a kitchen? (See "Lighting for Kitchen Activities" page 108)
2. What are the key activities to light in a bathroom? (See "Lighting for Bathroom Activities" pages 114–123)
3. Identify examples of spatial, emotional, and aesthetic visual experiences. (See "Spatial Experience" pages 124–125, "Emotional Experience" pages 125–127, and Aesthetic Experience" page 128)
4. Discuss two issues that affect the integration of electric light and daylight. (See "Daylight Integration" pages 128–129)
5. Discuss how recessed lighting in the ceiling differs from concealed lighting in a cove, in terms of both lighting effect and architectural issues. (See "Architectural Integration" pages 128–129)

Choosing Electric Light Sources

Light sources do the work when it comes to lighting. Incandescent—our oldest electric light source—as well as newer fluorescent and newest light-emitting diode (LED) technologies serve as the primary light sources for residential kitchens and bath applications. Each source is different, but they all share key attributes. These include lumen output, power, efficacy, distribution, color, size, life, and control.

Learning Objective 1: Recognize the primary light sources for kitchens and baths.

Learning Objective 2: Explain the principal attributes of light sources.

Learning Objective 3: Distinguish the differences in the principal attributes.

Learning Objective 4: Assess the importance of each attribute for lighting effectiveness.

LIGHT SOURCES FOR KITCHENS AND BATHS

In addition to daylight, discussed in Chapter 8, "Importance of Daylight," three electric light sources are commonly used in kitchens and baths:

1. Incandescent (including halogen and xenon)
2. Fluorescent (both linear and compact)
3. Light-emitting diodes (LEDs)

Each source produces light differently, leading to different sets of attributes and, from these, different benefits and limitations.

A Brief History

Man-made lighting dates back thousands of years to flame sources (wood, cloth, and fiber, together with carbon-based material such as oil and tar). Until the nineteenth century, however, lighting was either largely unmanageable (daylight) or astonishingly expensive (and dirty and dangerous).

Two hundred years ago, the cost for 800 lumen-hours of light (the equivalent of operating a typical incandescent lamp for 1 hour) required about 4.5 hours of work. Compare that to today's lighting cost, at less than a quarter of 1 minute of labor.

Early incandescent lighting was demonstrated successfully in about 1880 in both England and the United States. However, the modern incandescent lamp did not appear until 1910, with the advent of the tungsten filament. Halogen technology (improved incandescent) was introduced in 1959, with significant development since then.

Fluorescent lamps were introduced in the late 1930s; commercial and industrial adoption was spurred by the war economy of the 1940s. Modern fluorescent lighting, including compact forms, dates from the 1980s and has continued to improve since then.

LED technology for general lighting has become practical only in the twenty-first century, although its use in signs and colored lighting effects is slightly older. Importantly, while incandescent and fluorescent technologies have largely reached developmental plateaus, LED technology continues to evolve at a rapid pace, offering the prospect of a pleasing, flexible, economical, and environmentally friendly electric light source.

ATTRIBUTES OF LIGHT SOURCES

All light sources share key attributes that determine which sources are most appropriate for which applications. No single light source will perform optimally for all applications.

Eight key attributes are:

1. Light output
2. Power
3. Efficacy
4. Light distribution
5. Color
6. Physical size
7. Life
8. Control

Light Output

In the past, residential light sources were measured by wattage—the power needed to operate. That is, we gauged the output of light by the input of power. Perhaps when incandescent lighting predominated, this approach worked. Today, we use a wide variety of light sources, and light output no longer correlates directly to power input.

Light output, technically luminous flux, is measured in lumens (lm). Output varies by type of light source and also within the category of each source.

- A source with only a little light output (e.g., suitable for decorative applications) might provide 100 lm.
- A source more suitable for brighter indoor applications might emit 500 to 3000 lm. (These ranges are just indicative.) Several such sources might be used in a single luminaire.
- Sources with still higher output are widely used in exterior, commercial, and industrial applications. However, this is too much light in one place for residential use.

Output declines over time (lumen depreciation), with the rate of decline varying by light source. Halogen sources might lose only 5 to 10 percent of their lumen output over their short lives. LED sources last much longer and experience lumen depreciation of 30 percent. (There is more to this story, as we will see shortly.) Fluorescent sources combine excellent lumen maintenance (the opposite of depreciation) and long life, which produces consistent light and easy maintenance over long periods of time.

Lighting Facts Per Bulb

Brightness	820 lumens

Estimated Yearly Energy Cost $7.23
Based on 3 hrs/day, 11¢/kWh
Cost depends on rates and use

Life
Based on 3 hrs/day   **1.4 years**

Light Appearance
Warm   Cool
2700 K

Energy Used	60 watts

Lighting Facts Per Bulb

Brightness	870 lumens

Estimated Yearly Energy Cost $1.57
Based on 3 hrs/day, 11¢/kWh
Cost depends on rates and use

Life
Based on 3 hrs/day   **5.5 years**

Light Appearance
Warm   Cool
2700 K

Energy Used	13 watts

Contains Mercury
For more on clean up and safe
disposal, visit epa.gov/cfl.

Front   Back   New Back Label for Bulbs
New Light Bulb Label   Containing Mercury

FIGURE 10.1 Lighting facts label

Lumen ratings for common incandescent, LED, and compact fluorescent lamps (CFL) are printed on their packages. Look for the Lighting Facts label, which provides lumen output along with color, wattage, energy used, and life (see Figure 10.1).

For other lamps, notably linear fluorescent and low voltage types, you typically need to consult a manufacturer's catalog or website data. Note that lumen depreciation is typically available only from the catalog or website.

Power

"Power" refers to the *wattage* required to operate the light source. All electric light sources require power. Energy consumption is simply the power used over the time that the light source operates.

> ## Energy equals power consumed over time.
>
> $E = W \times T$ To calculate this, you would take the wattage of the light source and multiply it by the amount of time (usually expressed in hours).

It is important to recognize that the power required does not represent the light emitted. The relationship between light and power (discussed in "Efficacy," page 140) varies significantly among different light source technologies.

Among lamps of the same technology, however, differences in wattage generally translate into differences in light output (and sometimes physical size). As we increase our use of LED lighting, however, even that will not be true.

For a century, people have identified lamps—and the quantity of light desired—by their wattage: "I need a 100-watt bulb; the 60-watt wasn't bright enough." Times and technology have changed.

Today, we need to speak more accurately.

- If you want more light, ask for it! "I need more light; the 400-lumen lamp was not adequate, I need an 800-lumen model." As discussed, lamp packages now carry this information (as a matter of law).
- If you want to use less energy, you can ask for lower wattage. "Do you have a lower-wattage source with about the same lumen output?"
- You can also combine these ideas: "I don't need this much light and would like to save some energy. Do you have a lamp with lower lumen rating and less wattage?"

Efficacy

"Luminous efficacy" relates the output of the light source to the power required to operate it. We use the term "efficacy" (derived from "effect" and "effectiveness") to compare outputs (light) and inputs (power) that are different. Luminous efficacy is expressed in lumens (light) per watt (power) and is typically abbreviated as lm/W or LPW. Efficacy varies among light sources and (to a lesser degree) within the family of light sources.

Since energy typically represents the largest component of operating cost, comparing the efficacy of different light sources suggests which sources will be most economical.

Light Distribution

Light sources may distribute light in an omnidirectional way (think of a typical household lightbulb), in a more cylindrical pattern (linear fluorescent source), or focused into a directional beam (reflectorized lamp). For directional sources, we often measure the intensity of the source (candela or candlepower) in the direction of the beam and the angle of the beam spread (in degrees) (see Figure 10.2).

With directional light sources, a "spot" distribution is concentrated into a tight beam (less than 15 degrees in beam spread). A "flood" distribution is broader (36–40 degrees is typical). A narrow flood falls between the two. A wide flood is about 60 degrees, and some very diffused sources are wider still. For details, such as the beam angle, you may need to consult the manufacturer's catalog or website.

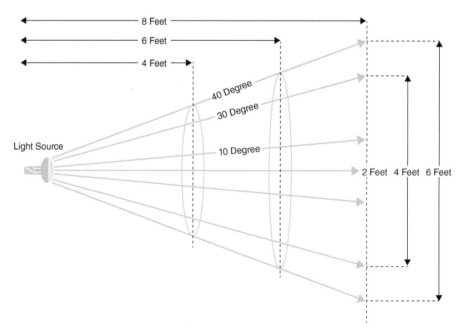

FIGURE 10.2 Diagram of beam spreads

Whatever the fundamental distribution from the source, the luminaire will modify it to a greater or lesser degree depending on its optics.

Color

Color rendering ability and color appearance are critical considerations in light source selection—particularly in kitchen and bath applications. We first discussed this topic in Chapter 2, "Seeing Materials," and Chapter 3, "Seeing the Space and Each Other." Now it is time to quantify these attributes so that we can measure light sources, compare the results, and select those most appropriate to the application.

Color Temperature

Color temperature expresses the appearance of the light itself—its tonality, "whiteness," or feeling. In an effort to make this concept accessible to consumers, lamp package labels call this "light appearance." Recall that color temperature represents the amount of heat needed to make metal glow to that color.

Of course, we do not heat an actual piece of metal to get the color temperature; we use a computer model of a perfect thermal radiator, known as a black body.

Only sources that produce light by heating metal (incandescent) can be said to have a color temperature. Other sources, such as fluorescent and LED, have a *correlated* color temperature (abbreviated CCT). That is, their color appearance has been mathematically related to the black body model.

Under this method of calculating color temperature, two light sources that are spectrally different may correlate to the same CCT value. Moreover, the measurement standards for CCT allow for a wide variance. That is, a source with a nominal CCT of 3000 Kelvins (K) may fall within a range of +/– 250 K. (This is established under the American National Standards Institute working with the lighting industry.) In other words, sources with the same CCT may not appear the same. This is a problem best resolved by observing each source rather than relying on the CCT rating alone. Checking the manufacturer's stated tolerance will also help determine the consistency of color from the lamp or luminaire.

Because color temperature varies so greatly, Table 10.1 shows recommended Kelvin temperatures, depending on the usage in both residential and commercial spaces.

TABLE 10.1 Color Temperature in Application

Color Temperature in Kelvin	Common Residential Use	Common Commercial Use
2200–2500	Dimmed incandescent in social areas	Dimmed incandescent in hospitality applications
2700	Typical incandescent lamps, not dimmed; most CFL; warm-colored LED lamps	Warm-colored CFL and LED lamps in hospitality and some upscale retail applications
2800–3100	Typical halogen lamps, not dimmed	Halogen for display lighting in museums and high-end retail applications
3000	Linear fluorescent, most appropriate for facial rendering and general lighting for work areas; also available for LED and CFL	Linear fluorescent, LED, and ceramic metal halide for retail and hospitality applications
3500	Linear fluorescent for utility areas	Linear fluorescent, CFL, and LED for office and educational applications
4000	Linear fluorescent for utility areas	Linear fluorescent, CFL, and LED for office and educational applications
5000	Probably too "cool" for most North American homes	Linear fluorescent, CFL, and LED for office and educational applications

- Warm-toned sources (2700–3000 K) are the most widely used in North American homes. Even warmer color (dimmed incandescent reaches about 2200 K) can be pleasing in some social settings but often feels yellow and "muddy" when applied as general lighting for working areas. These light sources are also used in commercial hospitality and high-end retail applications.
- Light sources emitting a cooler tone of light have a higher color temperature (3500–4000 K). These are commonly used in commercial office, education, and general retail applications.
- So-called daylight lamps have a noticeably cool color temperature (5000–6500 K). While this is well within the range of daylight, electric light of this color feels chilly and unnatural in most North American homes (but is widely used elsewhere in the world).
- The color of light can also be used to maximize the interior color palette. In rooms with cooler colors, 3000 K will maximize the black, white, grays, and blues. Using 2700 K in warmer rooms will bring out the browns in wood and keep yellows, beige neutrals, and orange comfortable.

Color Rendering

Color rendering represents the light's ability to interpret color. Good color rendering helps us to look our best and to present spaces, furnishings, and food in the best light. But how do we measure what is good?

The most widely used metric for color rendering is the color rendering index (CRI), first introduced about 50 years ago (see Figure 10.3). The CRI rates light sources against a reference source of the same color temperature, using a palette of eight pigment samples. The highest score on the CRI scale is 100, which represents a perfect match of the light source to the reference for the specific samples.

The sample material colors, while developed empirically, have relatively low saturation and obviously do not represent *all* materials, or those in any specific application.

The reference source for light sources with a color temperature of less than 5000 K (i.e., virtually all residential and most commercial lighting) is a computer model of the glowing metal we discussed in Chapter 3. The model is adjusted to produce a reference at the same color temperature as the specific light source being rated. Effectively, this is an incandescent source with its glowing tungsten metal filament.

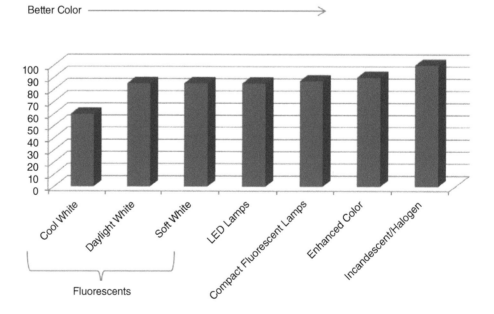

FIGURE 10.3 Color rendering index

For light sources about 5000 K, we use a model of daylight.

Incandescent and halogen sources provide a CRI of 100 (with a few exceptions) and warm color temperature. LED and fluorescent sources offer a much wider choice, with the attendant advantages and complexities of multiple options.

Using Color Rendering

A few key points about CRI:

- A higher CRI is generally better than a lower CRI. A CRI of at least 80 is recommended for residential applications.
- Comparing the CRI across light sources of *different* color temperatures is problematic since they are rated against different references. Evaluating the sources by direct observation is better.
- For warm-colored sources (those mostly used in North American homes), the CRI favors lamps with strong red content. Red-rich pigments are rendered vibrantly by such high CRI sources. Conversely, these high-CRI lamps tend to lack blue content; blue-rich pigments are therefore rendered less vibrantly.
- Since the reference model closely matches a tungsten light source, most incandescent and halogen lamps rate 100 on the CRI scale. This suggests that these sources offer perfect color rendering, which, of course, they do not. Their CRI rates 100 because they are— effectively—acting as a reference against themselves.
- LED and fluorescent sources are readily available with CRI of 80 or higher. These technologies can attain a CRI over 90 but generally sacrifice some efficacy to do so, and they typically cost more than similar lamps with a CRI of 80 to 90.
- Most important, judge for yourself by looking at light sources and your project's color palette before choosing. Do not forget to evaluate skin tones and even foods under the light sources.

Physical Size

In lighting, size matters. We have already discussed light sources in terms of the *amount* of light emitted (lumens) and power used (watts). Both of these attributes suggest size, and we often use the terms "bigger" and "smaller" to refer to them. Here we are using size to describe the *physical* properties of a light source.

- Smaller light sources permit the use of smaller luminaires, which often means a more attractive model. Smaller sources also permit more *precise optical control* of the light. Think of a small LED light source.
- Larger light sources provide a *more diffuse* (and often more comfortable) source of light. Think of a linear fluorescent source.
- As alternate light sources become more popular, the physical shapes of lighting likely will morph into new standards of normalcy.

Life

The life of a light source affects how frequently it will need to be replaced, with the resultant cost and inconvenience. The more you use the light source (hours per year), the more important its life is.

The life of a light source is rated in hours of operation. To understand life in calendar terms (days, months, years), you need to consider the hours of usage. For example, a lamp rated at 1,000 hours—an ordinary incandescent lamp—and operating three hours per day (1,095 hours = 365 days × 3 hours per day) would have a rated life of a little less than a year. For the same usage, a lamp rated at 25,000—a comparable LED lamp—would have a rated life of more than 22 years.

All manufacturing produces variation in the product. Some variances are large; some, small. Expressing this in terms of lamp life, we use an average and speak of the *rated average life* of a lamp.

To avoid the distorting impact of a few lamps that might last a very long time, we use a *median* value. The median is the middle value of a sample. Half the sample experiences a shorter life than the median, and half experiences a longer life.

Considering just a single lamp, you cannot know its actual life, which may be longer than the rated average or shorter. While this may sound rather complicated, it is typical of many industrial products. While we often say, "This lamp will last 25,000 hours (or 22 years)," we should remember that rated average life is a statistical prediction and not the actual life of any specific lamp.

What Is Average?

Average is a statistical measure; there are several commonly used types of averages. What often comes to mind first is the *mean*, as in the average age of a population. (Add up everyone's age and divide by the size of the population.)

What is the average number of lamps in a home? Take a sample of homes, add up all the lamps, and divide by the number of homes in the sample. Estimates by the US Department of Energy published in 2013 (using more sophisticated data techniques) put the average number of lamps of all types in US homes at 67.

Other attributes of light sources—lumen output, power, color—also vary. But unlike life, these attributes can be measured at the time of production, and their variance can be tightly controlled, either in the production or in postproduction quality control.

Control

How does a light source start up? Can it dim? These are the two key issues regarding control.

For residential usage—where we want light "on demand"—we need light sources that turn on and deliver light immediately (i.e., without a noticeable warm-up period). Incandescent and LED sources meet this requirement. Today's fluorescent sources also turn on right away. They take a little time to reach full light output, although this is rarely a problem.

Dimming allows us to adjust the quantity of light to suit different tasks, social activities, time of day, or personal preferences. Dimming makes a lighting system more flexible. And by dimming one luminaire to provide different lighting effects, it can do the work of two or more luminaires, which both simplifies the lighting installation and lowers its cost.

Here are several criteria for evaluating light sources in terms of their dimming capability:

- *Incandescent lamps grow warmer in color as they dim.* This can be important for creating the appropriate social atmosphere or a friendly night-light.
- *Fluorescent and most LED sources retain their color characteristics as they dim.* This can be valuable when simply changing the level of illumination to suit different tasks. A few newer LED products can warm up as they dim (to simulate incandescent dimming), but they must be specified to do so.
- *Dimmed sources should change their brightness smoothly—without visible flicker, hesitation, or skips—all the way down to a low level.* Incandescent lamps do this easily; fluorescent sources dim well, too. LED sources require careful consideration of source/dimmer compatibility.

Lighting Concepts

Take a trip to the local home improvement store. Visit the LED section of the lighting department and select a single product. Make the following observations:

1. What is the range of the lumen output offered by the LED product?
2. What is the range of color rendering and Kelvin temperature offered by the LED product?
3. What is the range of the wattage offered by the LED product?
4. Would the LED product be suitable for:
   a. Bathroom vanity area
   b. Under-cabinet task lighting in a kitchen
   c. General lighting in a walk-in pantry

SUMMARY

Light sources share key attributes that determine suitability for various applications. These include: light output, power, efficacy, distribution, color, size, life, and control. Incandescent is an old and familiar technique. Newer sources, such as fluorescent and—increasingly—LED, offer advantages in terms of higher efficacy and longer life. Color, size, and control vary among the sources and need to be considered when selecting the source.

REVIEW QUESTIONS

1. What are the eight attributes of light sources? (See "Attributes of Light Sources" page 138)
2. Discuss the difference between lumens and watts? (See "Light Output" and "Power" pages 138–140)
3. Discuss the difference between color rendering and color temperature. (See "Color" pages 141–143)
4. How is life measured for light sources? (See "Life" pages 143–144)
5. What are two key attributes of control in light sources? (See "Control" page 144)

Comparing Electric Light Sources

The three electric light sources appropriate for kitchen and bath spaces are incandescent, fluorescent, and light-emitting diode (LED). Each offers different lighting effects in terms of light output, distribution, color, and dimmability, and each has different limitations and costs. This chapter looks at each light source in turn by considering its effects and limitations.

> *Learning Objective 1: Identify the attributes, strengths, and weaknesses of incandescent, fluorescent, and LED sources.*
>
> *Learning Objective 2: Compare the light qualities of these sources.*
>
> *Learning Objective 3: Compare the light output, efficiency, and life of these sources.*
>
> *Learning Objective 4: Recognize the operational limitations of the three sources.*

LIGHT SOURCES FOR KITCHENS AND BATHS

In addition to daylight, discussed in Chapter 8, "Importance of Daylight," three electric light sources are commonly used in kitchens and baths:

1. Incandescent (including halogen and xenon)
2. Fluorescent (both linear and compact)
3. Light-emitting diodes (LEDs)

Each source produces light differently, leading to different attributes and, from these, different benefits and limitations.

Incandescent lamps create the charming glow or dramatic focal effects so critical for an appealing interior and pleasing atmosphere.

Fluorescent lamps produce a diffuse light more suitable for tasks and high levels of general illumination.

LED sources can provide both glow and focal beams of incandescent light and the effective task illumination of fluorescent light—with important limitations.

Well-designed kitchen and bath lighting can employ all three sources effectively, taking advantage of the distinctive attributes of each.

INCANDESCENT SOURCES

The word "incandescent" has the same Latin root as "candle," which is hardly surprising. Besides a technical definition, "glowing because of great heat," the word "incandescent" serves as a metaphor for such characteristics as passionate, brilliant, ardent, fervent, spirited, fiery, even intelligent and successful. Interestingly, it also shares its root, "white," with the word "candid."

Incandescent sources create light by passing electricity through a wire filament, heating it so that it glows, or incandesces. Electrical resistance, a property of the filament, results in heat and hence the light.

A Little Background

The ordinary incandescent lamp as we know it was more than a century in the making. Humphry Davy—a significant English scientist in the development of modern chemistry and electricity—demonstrated a dim and short-lived, battery-powered incandescent lamp in 1802.

Many demonstrations followed over the next three quarters of a century. Joseph Swan in England started research in the 1850s. In the United States, Thomas Edison took up the challenge in 1878. Hiram Maxim, an American born inventor who became a naturalized British subject, (subsequently, an inventor of machine guns) claimed to have invented the incandescent lamp, contended commercially and legally with Edison, and succeeded in lighting a New York City skyscraper in the 1880s, the first such installation in the United States. Until the introduction of tungsten filaments early in the twentieth century, incandescent lamps were still less efficient than gas lighting.

Ultimately, the competing parties resolved conflicting patents and technologies, and something like our familiar-looking incandescent lighting industry emerged.

Attributes of Incandescent Light Sources

Broadly speaking, people like familiar incandescent lighting because of the quality of light: warm in color; easy to shape into the desired beam, wash, or glow; and readily dimmed. Yet incandescent lighting comes at a high cost in electricity, maintenance, and environmental impact.

Reasons Not to Choose Incandescent Sources

Here are five significant reasons that people choose fluorescent and LED sources rather than the familiar incandescent ones.

1. *Incandescent sources are inefficient.* More than 90 percent of the electric energy used converts into heat, not light. Incandescent sources therefore require three to five times as much energy to create an equal amount of light when compared to fluorescent and LED sources.
2. *Incandescent lamps heat up a brightly lighted room.* Obviously, this results from the inefficiency of making light by slowly burning up a wire. Dealing with heat from light can be a problem, particularly in kitchens and bathrooms.
3. *Incandescent lamps have short life spans.* Incandescent sources do not last very long compared to fluorescent and LED sources. Ordinary household lamps typically have a rated average life of 1,000 hours—a little less than a year in the calculations of the US Department of Energy. Operating during daylight hours, the average lamp might last less than six months. Given the 67 lamps in a typical American home, short life means nearly continuous maintenance of the lighting systems.
4. *Incandescent sources are expensive.* When considering both energy usage and short life, incandescent sources cost more than the alternatives. Even if incandescent lamps were free, using them would cost more than buying and using LED or fluorescent lamps.

5. *Incandescent sources have an environmental cost.* Because of its inefficient use of energy, incandescent lighting imposes more environmental cost than other light sources.

Reasons to Choose Incandescent Sources

Here are five significant reasons that people continue to use incandescent lighting despite its operational inefficiencies, high cost, and environment impact.

1. *Color and texture.* Incandescent sources provide the color and texture of light most people have grown up with and feel most comfortable with. Incandescent sources are rich in long-wavelength (red) energy and weak in short-wavelength (blue) energy. They vary between about 2,500 Kelvins (K) and 3,000 K, with decorative lamps at the yellower end of the spectrum and halogen at the crisper end.
2. *Shape.* You can readily shape incandescent light into concentrated beams, soft washes, crisp grazing, and gently glowing pools. This optical control derives from the small size of the filament, which enables the light output to be gathered and redirected easily. For this reason, incandescent sources are sometimes called *point sources* (while fluorescent is an area source).
3. *Effectiveness.* With the ability to shape light, incandescent lighting may be the most energy-effective light source at delivering the desired lighting result.
4. *Dimmability.* You can dim incandescent sources easily and cost effectively. Dimmed incandescent can achieve remarkably low intensity with smooth gradients from setting to setting. And, as it dims, incandescent light grows pleasingly warm—almost like candlelight.
5. *Practical, flexible, and low cost.* After a century of development, incandescent lighting is highly practical. Many thousands of lighting fixtures accept incandescent sources, often including multiple sources in the same fixture, which offers considerable flexibility. And the initial cost of incandescent lighting is generally less than that of other sources. Many consumers continue to select incandescent light sources because they are most familiar with the terminology of incandescent lamps.

Construction

Incandescent lamps consist of a tungsten filament enclosed by a glass "bulb." The bulb is filled with gas that retards evaporation of the filament and so increases lamp life. Argon gas is used in most simple incandescent lamps over 25 watts (W). Bromine gas, a halogen element, is used in the more advanced lamps in the halogen incandescent family. The most common bases for incandescent lamps use a threaded screw shell (see Figure 11.1) for line voltage or a bi-pin connection for low-voltage applications.

Imagine electrons flowing through wire. Thinner wire constricts the flow, which raises resistance, heats the wire, and increases the flow of light. Of course, the hotter the wire, the faster it evaporates and breaks. Conversely, thicker wire eases the flow of electrons, which reduces resistance, heat, and light.

Nomenclature

Figure 11.2 is a common household bulb. The nomenclature for incandescent lamps is straightforward:

43A19/suffixes

43 Wattage is shown first. This lamp is the incandescent replacement for the ubiquitous 60A19, the production and import of which in North America ceased in January 2014 (more on this later). It is actually a halogen lamp (although that is not part of its nomenclature).

A Shape is designated by a letter. A stands for arbitrary. Other common types are shown in Table 11.1.

19 The size of the lamp is designated by its diameter at the largest point. This is denominated in eighths of an inch. Here 19 means 19/8s of an inch, or 23/8 inches in diameter (see Figure 11.3).

FIGURE 11.1 Screw Shell Incandescent Lamp Bases
Courtesy of Vvoe Vale

FIGURE 11.2 Common household bulb

TABLE 11.1 Common Incandescent Lamp Shapes

Shape	Meaning	Application	Image
A	Arbitrary Sometimes referred to as GLS for General Lighting Service	General Soft lighting or sparkle from decorative luminaires, portables, or downlights	43A19
G B F	Globe Decorative Flame	General Soft lighting for sparkle from decorative luminaires or exposed application	25G25 B10
T	Tubular	General Soft lighting for sparkle from decorative luminaires, portables, task lighting, or exposed application	T3 T4 T10
R BR	Reflector Bulge Reflector	Directional Soft-edged flood lighting, most commonly from recessed downlights	R20 BR30
PAR	Parabolic Aluminum Reflector	Directional Sharp-edged spot or flood lighting, most commonly from accent lights and recessed downlights	PAR16 PAR30L
MR	Multi-Reflector	Directional Sharp-edged spot or flood lighting from accent lights or recessed downlights, decorative or portables	MR16
	Festoon	General Soft wash, typically used in linear low-voltage systems	Festoon

Suffixes

The suffix may designate the finish of the lamp (clear or frosted), lamp type (halogen or "ordinary" incandescent), beam spread (for directional lamps), or the voltage for which the lamp is rated. Lamps may use several suffix designations.

Note what the incandescent lamp nomenclature does *not* convey:

- *Source type.* Incandescent is understood.
- *Light output.* The lamp nomenclature does not tell you how much light is emitted. You find that out from manufacturer's information or the lamp package.
- *Lamp base.* The lamp nomenclature does not tell you whether the lamp is compatible with your fixture's lamp holder. You find that out from manufacturer's information. (See Table 11.2.)

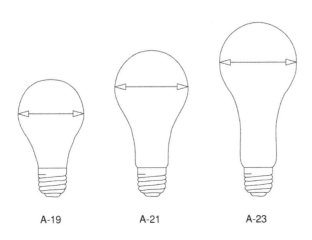

A-19   A-21   A-23

FIGURE 11.3 Common incandescent bulb sizes

TABLE 11.2 Common Incandescent Lamp Bases

Type	Lamps	Application
Medium (screw shell)	A, G, T, R/BR, PAR	Most common base for incandescent lamps; with appropriate fixture design, the ubiquity of the medium base allows for a very flexible lamp choice
Candelabra (screw shell)	Decorative types	Decorative luminaires with low wattage per socket and limited need for flexibility in application
Mini-Candelabra (mini-can)	T	Decorative luminaires, projector, or downlights using higher-wattage halogen lamps
Bi-Pin GU4 GU5.3 GY6.35 GY8.6 G4	MR, T All but GY 8.6 are for low-voltage lamps	Common for low-voltage lamps There are several types of bi-pin bases and lamp holders, and they are *not* interchangeable
GU10	120V MR lamps	Downlights and track luminaires using 120V MR16 lamps (generally for lower cost than 12V luminaires)

- *Overall length.* You know the diameter but need to look up the length on manufacturers information.

Here is another useful lamp:

40PAR30L/HIR/NFL25

40 This 40-W lamp is the current version of a 75-W lamp introduced a quarter of a century ago.

PAR The parabolic shape of the lamp's aluminized reflector concentrates light from the filament into well-defined beam (see Figure 11.4).

FIGURE 11.4 PAR30L lamp

30L The diameter of the lamp is 3¾ inches. (Do the math.) The "L" designates this as a long-neck lamp, which is used in recessed downlights. The short-neck PAR30S is used more commonly in track lighting, where its compact form permits the design of smaller and more attractive fixtures.

HIR Designates halogen infrared (discussed later) for one manufacturer. Other manufacturers use similar but different catalog nomenclature.

NFL25 Narrow flood 25-degree beam. This is standard nomenclature.

Halogen Technology

Most incandescent lamps today use halogen technology. This includes A lamps, PAR lamps, MR lamps, and assorted other types. Halogen lamps take their name from the family of gases they contain. (Bromine is most common.)

Unlike the inert argon gas in ordinary incandescent lamps, bromine interacts with the tungsten evaporating from the filament of the lamp, capturing it before it is deposited on the outer envelope of the lamp and redepositing it on the filament. For the "halogen cycle" to work properly, the bromine gas and tungsten filament need to operate at a high internal temperature and must be contained by a strong quartz glass enclosure (see Figure 11.5).

Benefits of Halogen Technology

The halogen provides several benefits compared to ordinary incandescent operation:

- Longer life due to the redeposition of tungsten onto the filament
- Brighter, whiter light due to a higher internal temperature
- Higher efficacy due to the higher internal temperature

Notwithstanding the benefits, halogen lamps are still less efficient, and experience shorter life, than fluorescent or LED sources.

Infrared Coating

Incandescent lamps produce light by heating up a filament until it glows; at the same time, 90 percent of the electricity running through the lamp winds up as waste heat in the form of infrared energy. Does this suggest a method of making incandescent lamps more efficient?

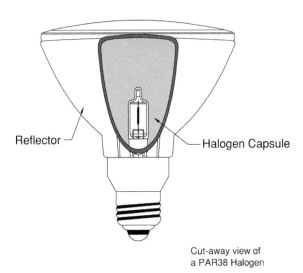

Reflector —   — Halogen Capsule

Cut-away view of
a PAR38 Halogen

FIGURE 11.5 Halogen capsule in lamp

Infrared films coating a halogen capsule permit visible wavelengths of energy to pass through as light while reflecting longer, infrared wavelengths back onto the filament, where they replace some of the electricity needed to heat the filament to incandescence.

Using this technology, infrared-coated halogen lamps *save about one-third of the energy* of simple halogen lamps. Today, most PAR lamps use infrared-coated halogen technology. Infrared-coated halogen lamps are denoted by such nomenclature as IR, HIR, and IRC.

Regulation of Incandescent Lamps

The US Department of Energy first regulated the energy efficiency of incandescent reflectorized lamps in 1992. The Energy Independence and Security Act (EISA) of 2007 established efficiency and other requirements for most general service incandescent lamps, creating widespread—but ungrounded—fears of bans on incandescent lighting.

In fact, over the last two decades, the efficiency and life of incandescent lamping has increased, in some cases significantly. Table 11.3 shows the improvements, following the final phase-in of EISA 2007 requirements in 2014 and the most recent updates to reflector lamp requirements.

Looking ahead, EISA requirements that take effect in 2020, appear impractical with current incandescent and halogen technology. By that time, however, both the performance and the price of LED lighting may have improved to the point that the market will be satisfied to relegate incandescent lighting to decorative effects.

Low-Voltage Lighting

Ordinary household electrical circuits distribute electricity at about 120 volts. Low-voltage lighting typically operates at 6 to 24 volts. Lower voltage permits smaller filaments in the light source and, from this, smaller lamps and luminaires.

The most common incandescent and halogen low-voltage lamps operate at 12 volts; 6- and 24-volt types are also available. These lamps need a transformer to reduce the voltage from 120 volts at the line to the voltage for which the lamp is designed.

Transformers reduce the voltage using either magnetic iron and copper materials (older types) or electronic components (newer types). While the type of transformer may be a question of industrial design for the luminaire manufacturer, it affects the choice of dimming control. (More on this in Chapter 12, "Light Fixtures," and Chapter 13, "Lighting Controls.")

TABLE 11.3 Replacements for Popular Incandescent Lamps*

Old Lamp	Current Lamp	Improvement	Life
100A19	72A19 (Halogen)	28% lower wattage	33% longer life
60A19	43A19 (Halogen)	28% lower wattage	Same life
90PAR38/H[1]	55PAR38 (HIR)	38% lower wattage	46% longer life
50MR16[2]	35MR16 (HIR)	35% lower wattage	100% longer life
75R30[3]	65BR30	13 lower wattage	Same life

* Replacement typically is based on delivering comparable lighting performance. In the case of general service lamps, equivalence is based on lumen output. In the case of reflectorized lamps, equivalence is based on intensity of the beam (maximum beam candela value). Values shown here are typical. Higher-performance products are available in many cases.
[1] 90-W *halogen* lamps replaced earlier 150 incandescent PAR lamps beginning in 1992.
[2] MR lamps are not yet regulated. Both halogen and superior halogen IR lamps are available.
[3] BR lamps replaced R lamps in 1992. They have not been regulated since.

The transformer may be housed in the luminaire, where it drives a single lamp, typical of recessed downlights and track lighting. Alternatively, the transformer may power a string of low-voltage lamps, which is common in strip lights used in coves or under cabinets.

Benefits of Low-Voltage Technology

Low-voltage technology offers three benefits:

1. Smaller luminaires, due to the smaller size of the lamp, including recessed downlights and strip lights
2. Excellent optical control, due to the small size of the light source filament
3. Higher efficiency, especially compared to incandescent lighting at comparable wattage

Since many early halogen lamps were low voltage, the similar benefits of these two technologies sometimes are conflated.

Dimming

As we see when we look at controls in detail, incandescent sources dim easily and with relatively low cost. Incandescent lamps can:

- *Dim down to very low levels.* Exposed filaments create elegantly visible "signatures" at such a low levels.
- *Create a warm, intimate environment.* As they dim, incandescent sources become amber, producing a color temperature around 2200 K. Of course, this means that dimming changes both the color and volume of light in a space. You cannot separate the two effects.
- *Last much longer.* Dimming a typical incandescent lamp by 20 percent increase its life by *four* times. (How long a dimmed lamp lasts depends mostly on much it is dimmed.)

Dimming does have its limitations, however. Dimming significantly reduces the light output of an incandescent lamp, so that 20 percent dimming costs more than 25 percent of the light output. Thus, dimmed lamp efficiency falls. Dimming incandescent lamps also changes the output color of light, placing more amber/yellow into the space.

Incandescent Lamp Options

The hundreds of different incandescent lamps fall into several broad families.

- *General service—line voltage.* The versatile A lamp in several sizes is the most important product here.
- *Decorative—line voltage.* G lamps for bathroom strip lights and flame shapes for chandeliers predominate.
- *Directional—line voltage.* BR and PAR lamps in various sizes and beam spreads are useful for downlighting (use flood lamp (FL) and wide flood lamp (WFL) distribution) and accent lighting (choose distribution based on objects being lighted).
- *Directional—low voltage.* The MR16 lamp, derived from slide projectors in the 1970s, offers the most choice of intensity and beam spread for downlighting (use FL and WFL distributions) and accent lighting (choose distribution based on objects being lighted).
- *Nondirectional—line and low voltage.* These lamps find use in decorative luminaires, portables, and linear cove and task lighting.

Table 11.4 shows a selection of typical incandescent lamps with lumen (lm) values and life ratings. The shape and wattage are given by the lamp type.

Looking at the column for efficacy, note how it is higher for halogen and low-voltage lamps than for ordinary incandescent and 120V lamps and how, within any type of lamp, efficacy increases with wattage. Table 11.5 is a summary of the attributes of incandescent light sources.

TABLE 11.4 Typical Incandescent Lamps with Lumen Values and Life Ratings

Lamp	Lumens	Efficacy (lm per watt)	Life (hours)	Notes
29A19 (Halogen)	400	13.8	1000	White or clear
43A19 (Halogen)	750	17.4	1000	White or clear
72A19 (Halogen)	1490	20.7	1000	White or clear
150A21	2620	17.5	750	Inside frost
25G16.5 (Halogen)	270	10.8	2200	Clear, candelabra base
25G25	235	9.4	2000	Clear
40G25	415	10.4	2000	White
25BA11 (Halogen)	280	11.2	2750	Clear, candelabra base
40 BA10.5	300	7.5	2000	Clear, candelabra
40R20	380	9.5	2000	
65BR30	620	9.5	2000	Flood
50PAR20 (Halogen)	570	11.4	3000	SP10 and FL25
50PAR30 (Hal/IR)	850	17.0	3000	SP10 and FL25
60PAR38 (Hal/IR)	1070	17.8	3000	SP10 and FL25
50PAR30 (Hal/IR +)	900	18.0	4400	SP10, FL25, WFL40
55PAR38 (Hal/IR +)	1100	20.0	4400	SP10, FL25, WFL40
20MR16 (Hal)	240	12.0	3000	SP10, FL36, 12V
35MR16 (Hal)	540	15.4	3000	FL36, 12V
50MR16 (Hal)	800	16.0	3000	SP10, NFL24, FL36, 12V
20MR16 (Hal/IR)	320	16.0	5000	SP8, FL36, 12V
35MR16 (Hal/IR)	750	21.4	5000	SP8, NFL24, FL36, WFL60, 12V
45MR16 (Hal/IR)	1050	23.3	5000	SP8, NFL24, FL36, WFL60,12V
25MR16/GU10 (Hal)	160	6.4	2000	FL25, 120V
35MR16/GU10 (Hal)	265	7.6	2000	FL25, 120V
50MR16/GU10 (Hal)	700	14.0	2000	FL25, 120V
50T4 (Hal)	500	10.0	1000	Clear, mini-can base, 120V
100T4 (Hal)	1600	16.0	1000	Clear, mini-can base, 120V
150T4 (Hal)	2800	18.7	1000	Clear, mini-can base, 120V
35T4 (Hal/bi-pin)	400	11.4	2500	Clear, GY8.6 base, 120V
50T4 (Hal/bi-pin)	600	12.0	2500	Clear, GY8.6 base, 120V
10T3 (Hal/bi-pin)	140	14.0	2000	Clear, G4 base, 12V
20T3 (Hal/bi-pin)	320	16.0	2000	Clear, G4 base, 12V
35T4 (Hal/bi-pin)	600	17.1	2000	Clear, G6.35, 12V

Lamp data are from a major manufacturer; the listing is selective. Other manufacturers' offerings and data may be different.

TABLE 11.5 Summary of Attributes of Incandescent Sources

Light output	Wide range: 100–4000 lm for residential applications
	Good lumen maintenance (particularly for halogen)
Luminous efficacy	Low: 5–25 lm per watt depending on technology and wattage
Color	Warm: 2500-3000 K at 100 color rendering index (by definition)
Life	Short: 500–10,000 hours (longer when dimmed)
Auxiliary electrical devices	Transformers for low-voltage lamps
Types	Numerous, both general service and reflectorized
Operation	Instant on, no warm-up, easily dimmed

FLUORESCENT SOURCES

Fluorescent lamps, pursued by researchers from the beginning of the twentieth century, emerged in commercial reality in the 1930s. Fallingwater, Frank Lloyd Wright's iconic house designed for the Kaufmann family, was an early adopter (1939). The home, located in rural southwestern Pennsylvania, featured fluorescent lamps installed in niches around the windows and in coves over the bedroom cabinets.

Industrial mobilization for World War II accelerated the adoption of the technology. By the 1950s, fluorescent lighting was featured in architectural magazines featuring modern home design. Disappointment with the color and the flat quality of early fluorescent light largely relegated its use to lower-cost homes from the 1960s until the introduction of compact fluorescent lamps that featured better phosphor coatings and increased concern for energy efficiency.

Today, fluorescent lamps provide an important source of energy efficient, low-cost general lighting, with color and luminaire designs appealing enough for many, if not most, homeowners.

Fluorescent Technology

Compared to incandescent lamps, fluorescent sources offer the benefits of much higher efficacy; much longer life; and a soft, diffuse quality of light. However, fluorescent lighting can make a space feel flat and dull. Moreover, the poor color quality of some (typically older) fluorescent lamps is inappropriate for residential—and most commercial—applications.

How a Fluorescent Lamp Creates Light
Fluorescent technology is considerably more complex than incandescent (see Figure 11.6).

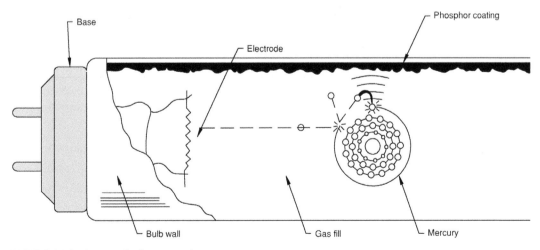

FIGURE 11.6 Diagram of a fluorescent lamp

Fluorescent lamps create light using a three-step process:

1. An electric arc (flow of electrons through a gas or vacuum) is struck between two cathodes inside a glass tube. The cathodes are coated with a barium emission material to assist with electron emission during the start-up.
2. The electrons stimulate a small quantity of mercury vapor within the gas filling the tube. The energized mercury discharges ultraviolet (UV) light.
3. The UV discharge reaches the glass tube, which is painted on the inside with a phosphor compound. Phosphors (light-reactive minerals), convert some of the UV radiation to visible light.

The *lumen output* of the lamp depends primarily on its length, the phosphor coating, the gas fill, and the current flowing through the lamp. *Color* depends on the phosphor mix.

Most fluorescent lamps turn on instantly but require some time to warm up to full output. (This can be a problem with compact lamps and cold environments.)

Ballasts

In addition to the glass tube with its cathodes, phosphors, mercury, fill gas, and base, a fluorescent system requires an auxiliary device, called a *ballast*. The ballast provides the correct starting and operating electrical input and must be compatible with the lamp it is driving. As we discuss shortly, fluorescent lamps can be dimmed, with a dimming ballast and appropriate dimming control.

Early ballasts used magnetic technology, with an iron core and copper windings around it. These ballasts might hum and could create noticeable flicker. Modern ballasts mostly use electronic components. Their high-frequency operation is more efficient, quiet, and flicker-free.

Families of Fluorescent Lamps

Fluorescent technology is divided into three broad families of lamps, each with applications in kitchens and bath areas (see Table 11.6).

1. *Linear lamps.* These are the familiar fluorescent lamps with a two-pin base at either end. Linear fluorescent lamps provide most of the general lighting in commercial and industrial applications. They can be quite useful in residential applications as well, if they have the appropriate color.
2. *Compact nonintegrated lamps.* Tubes of these lamps are folded back onto a single base, typically with four pins. They deliver the energy and maintenance benefits of fluorescent technology in a compact form. These lamps are often called pin-base (or plug-in) compact fluorescents (CFLs) to distinguish them from the integrated type discussed next.

TABLE 11.6 Families of Fluorescent Lamps

Family	Applications	Varieties
Linear	General area lighting Task lighting Cove lighting	Output/length Tube diameter Color
Compact nonintegrated	Downlighting Decorative luminaires	Output/wattage Tube configuration Color
Compact integrated	Replacement of incandescent in downlights and decorative luminaires	Output/wattage General/directional Color Dimmability

3. *Compact integrated lamps.* These lamps integrate the ballast, which is housed in the base of the lamp, and an incandescent-type base. Commonly referred to as retrofit, screw-base, or simply CFLs, these lamps are designed for use in incandescent fixtures and portables.

Nomenclature

The nomenclature for linear lamps is straightforward; for compact lamps, the nomenclature is much less clear.

Linear Lamps

Linear fluorescent lamps follow a simple, easy-to-use nomenclature (see Figure 11.7).

F32T8/841/ES

F F for fluorescent

32 32 watts or designed for a 32-W ballast. Wattage varies with the length of the lamp. As we will see, the *actual* power used by the lamp depends on the ballast.

T8 Tubular design that is eight eights, or one inch in diameter. The most popular diameter today is T8. (T12 used to predominate.) The other important size is T5, which is just 5/8s–inch in diameter.

841 841 designates lamp color. 8 = color rendering index (CRI) between 80 and 89. 41 designates color temperature of 4100. All manufacturers use this nomenclature. Familiar names such as Warm White, Soft White, Cool White, and Daylight are still used in consumer packaging. Check the technical designation on the lamp or package to be sure what the name really means.

ES A manufacturer's own suffix for energy saving. Other suffixes might indicate a long-life version. Each manufacturer uses slightly different terms.

As with incandescent lamps, the catalog number of the lamp does *not* tell you some important information.

- *Length.* F32T8 lamps are 48 inches long; F25T8 lamps are 36 inches long. Wattage and tube diameter determine linear lamp length. Note that nominal 4-foot T5 lamps measure 46 inches.

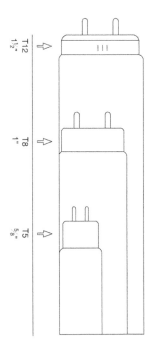

FIGURE 11.7 Diagram of T12, T8, and T5 diameters

- *Light output.* The lamp nomenclature does not tell you how much light is emitted. You find that out from manufacturer's information or the lamp package.
- *Lamp base.* T8 and older T12 lamps use a medium bi-pin base and the same lamp holders. T5 lamps use a miniature bi-pin base and lamp holder.

Compact Lamps

Here is the generic description of a typical pin-base compact fluorescent lamp, such as you might use in a recessed downlight. However, manufacturers often use their own trade names (PL, Biax, and Dulux are examples) and take their nomenclature from those trade names.

CFTR26W/GX24q/830

CFTR CFTR stands for compact fluorescent, triple tube (three sets of bent-over tubes on the base). CFQ (for quad) means two sets of bent-over tubes. CFT (twin) one set of bent-over tubes. CFT means a long, compact twin-tube (sometimes used for decorative fluorescent pendants).

26W Wattage. Since output depends on the length of the tube, triple-tube lamps are more compact than quad-tube lamps of equal wattage.

GX24q This is the base; some nonintegrated lamps of different wattages share the same base; others do not. When CFL lamps were introduced in the 1980s, they used two-pin bases and were operated off magnetic ballasts. Today's lamps mostly use four-pin bases and operate off electronic ballasts.

830 Compact lamps use the same color designations as linear lamps.

As with linear lamps, the catalog number of a compact lamp does not tell you either the length or the light output.

Nomenclature for integrated CFLs (retrofit or screw base, as seen in Figure 11.8) is even more mysterious since there are no established generic descriptions. Instead, most manufacturers use their own designations, and the lamps are commonly referred to by the names of the incandescent lamps they are intended to replace (with the exception of *spiral*, *spring*, *twister*, and other popular names).

FIGURE 11.8 CFL screw-base lamps

Character of the Light

Fluorescent light differs markedly from incandescent in most applications. Efforts to promote the replacement of incandescent with more efficient sources may overlook the essential differences.

The flat, diffuse quality of fluorescent serves well for task lighting (e.g., over a kitchen counter) and for bright general illumination (kitchens, workrooms, even bathrooms). However, fluorescent lighting diminishes the perception of form and texture, a concern in an elegant kitchen or bathroom.

When grooming at a mirror, the line between flat and flattering may be a little fuzzy. The smaller the light source, the more likely it is to create shadows at the eyes, nose, and chin (especially when the light source is located overhead). More diffuse light softens those shadows, making it easier to do the task and giving a softer appearance to your face.

Color

The color of fluorescent light depends on the phosphor mix that coats the inside of the lamp. Older fluorescent lamps generally used phosphors that created an unpleasant cast to the light, which was particularly bad for facial rendering. Better phosphors, while available, were considerably more costly and significantly reduced light output. Today's phosphors are refined from rare earth minerals and combine better color and better efficiency.

Called *tri-phosphor* technology, modern fluorescent recipes combine red, green, and blue light to produce a variety of white tones. Warm tones use more red; cool tones use more blue. With the right balance of phosphors, applied in a heavy coating, fluorescent lamps achieve a CRI over 80. Although lamps with a CRI over 90 can be found, they are more expensive and less energy efficient, and tend to be cool in color.

Compared to incandescent, fluorescent light is weaker in red but more balanced overall. Experience suggests that linear and pin-based CFLs offer better color than most integrated CFLs (notwithstanding similar CRI ratings).

In kitchen and bath applications, 830 fluorescent (80–89 CRI and 3000 K) generally provides the most pleasing color option among readily available lamps. Cooler 841 or even 850 lamps may be appropriate for utility spaces not visibly connected to the rest of the home.

The standards for color temperature permit a very wide variation. That is, lamps with the same Kelvin rating may look different. To avoid inconsistency of appearance, do not mix fluorescent lamps from different manufacturers. That is why when lamps are replaced in commercial spaces, entire rooms or floors are done to ensure color consistency.

Light Output

While there are other influences, light output varies with the length of the fluorescent tube (either linear or compact). This is quite different from incandescent lamps, which typically can deliver a range of light output from lamps of the same physical size. If you need more fluorescent light, you need longer lamps or more of them.

Unlike incandescent lamps, the light output of a fluorescent lamp is affected by the temperature around it.

T8 Linear Lamps

Ballasts also affect light output. For T8 lamps, a normal ballast generates about 88 percent of rated light output. (This is known as a .88 ballast factor). Wattage varies proportionately with light output. Lower- and higher-output (and wattage) systems are available, but these are mostly used in commercial applications.

As a rough rule of thumb, T8 fluorescent lamps produce roughly 550 to 700 lumens per foot. (Shorter lamps are a little less efficient and deliver a little less light per foot.)

Including the effect of the ballast, systems with good 4-foot fluorescent lamps deliver 90 to 100 lumens per watt (lm/W) (before losses within the fixture).

T5 Linear Lamps

Smaller-diameter linear fluorescent lamps fall into two categories: short lamps used for task lighting and longer versions for general applications. The longer versions, introduced from Europe in 1997, permit the use of narrower luminaires and can fit into small coves or behind short fascias.

T5 lamps perform about on a par with the best T8 lamps; their advantage is the ability to fit in smaller spaces. T5HO lamps, using proportionally more power, deliver about 1200 lumens per foot (a solution to very tight spaces).

Compact Lamps

The light output of compact lamps also depends on the length of the tube, although you would have to unbend the lamp to measure it. In practice, you can judge from the height of the lamp (tip to base). For more light, you generally need a larger luminaire.

In a luminaire, the bulky form of a CFL and the integrated ballast typically block some of its own light. As a result, the published equivalence of incandescent and CFL lamps often leads to less delivered light than expected. What is more, CFL lamps cannot replicate the concentrated beams of incandescent PAR lamps or the scale and beam control of MR16 lamps.

The efficacy of compact lamps ranges from 40 to 75 lm/W.

Light over Life

Like all other lamps, fluorescent lamps suffer reduced light output as they age; this is known as lumen depreciation. With today's high-quality phosphors, linear lamps lose only about 10 percent by the end of their lives; compact lamps lose about twice that.

Life

Fluorescent lamps, particularly linear lamps, last a long time—in a home as much as two decades, or more.

Each time the lamp starts and as long as it operates, some of the emission material evaporates. When the emitter is sufficiently depleted, the lamp no longer functions. Importantly, the more frequently the lamp starts, the faster the emitter evaporates. Ballasts also affect emitter depletion: Some start the lamp gently, prolonging life; others do the opposite.

To account for these operational characteristics, the life of fluorescent lamps is rated on a standard protocol. Using standard ballasts, lamps start and operate for 3 hours; then they are turned off for 20 minutes. The process repeats until the lamp no longer operates. The life rating is the median hours of operation in a large sample.

Under these conditions, linear lamps can start about 10,000 times, and compact lamps can start from 2000 to 5000 times. Under the worst operating conditions, lamps that were expected to last many years might fail in less than a year. Some of the gap would be the natural variation in manufacturing and some rapid on/off switching (think closet with a switch located at the door jamb).

Practical Light

With its high efficiency and long life, fluorescent lighting is easy on the pocketbook and easy to maintain. High efficiency and limited heat gain are particularly important where multiple lamps are combined with enclosed spaces (common in many bathrooms and kitchens).

Dimming

Using a dimming ballast and compatible control, fluorescent lamps can dim. Dimmable compact fluorescent screw-base lamps include a dimming ballast. Unless the lamp is marked dimmable and is controlled by a compatible dimmer, it will not behave properly.

Unlike incandescent sources, dimming fluorescent only lowers light output; life and color do not change. Moreover, it is difficult to dim fluorescents down to very low levels, particularly with compact lamps.

This means that dimming serves mostly to tune the fluorescent light level so that it is appropriate for different tasks, *not* for setting mood and atmosphere.

Mercury

All fluorescent lamps require small amounts of mercury for efficient operation. There is often objection to mercury due its toxic effects, but today's mercury content is as much as 80 percent lower than levels of 20 years ago. In any event, fluorescent lamps should be recycled to avoid contamination of water supplies through leakage from landfills.

If you break a lamp, carefully clean up the broken pieces using a dampened paper towel to collect the small particles, place the contents in a sealed plastic bag, and dispose at a hazardous waste collection site. Often municipalities offer this type waste collection at recycling centers. To avoid spreading the particles, do not vacuum or open a window until you have cleaned up the breakage.

To eliminate mercury altogether and preserve energy efficiency, consider LED lighting (discussed under "LED Sources").

Fluorescent Lamp Options

Fluorescent lighting offers the benefits of ample, diffuse general lighting with low installed and operating costs.

Table 11.7 shows selected fluorescent lamps from the three basic families. Note how, that energy-saver linear lamps offer improved efficacy, lower wattage, and longer life. Although they are used mostly for commercial applications, they offer better economics and sustainability for residences as well. You can also see the light output relative to the maximum overall length (MOL) of compact lamps and how covering a CFL (so you do not see the tubes) significantly affects its performance. Table 11.8 is a summary of the attributes of fluorescent sources.

TABLE 11.7 Selected Fluorescent Lamps

Lamp	Lumens[1]	Efficacy (lumens per watt)[2]	Life (hours)[3]	Notes
F17T8	1350	79	24,000	24" lamp
F25T8	2150	83	24,000	36" lamp
F32T8	2850	89	24,000	48" lamp
F32T8/ES	2500	100	32,000	48" lamp, 25 W, energy saver
F28T8/ES	2725	97	32,000	48" lamp, 28 W, energy saver
F14T5	1350	87	25.000	22" lamp
F21T5	2100	90	25,000	34" lamp
F28T5	2900	93	25,000	46" lamp
F28T5/ES	2900	104	35,000	46" lamp, 25 W, energy saver
CFQ13	900	62	12,000	5.2" MOL
CFQ18	1250	62	12,000	5.7" MOL
CFQ26	1800	62	12,000	6.5" MOL
CFTR18	1200	60	16,000	4.6" MOL
CFTR26	1800	62	16,000	5.0" MOL
CFTR32	2400	68	16,000	5.6" MOL
13-W spiral	900	69	12,000	Medium base, also GU24

(Continue)

Lamp	Lumens[1]	Efficacy (lumens per watt)[2]	Life (hours)[3]	Notes
18-W spiral	1300	72	12,000	Medium base, also GU24
23-W spiral	1600	70	12,000	Medium base, also GU24
5-W candle	215	43	8000	Covered for exposed use, candelabra base
9-W G25	500	56	8000	Covered for exposed use, medium base
14-W A19	800	57	8000	Covered for exposed use, medium base
15-W R30	750	50	8000	Reflector with extremely wide distribution
9-W fan	450	50	8000	Covered, medium base
20-W dim Sp	1250	63	10000	Dimmable, spiral
16-W dim R	630	39	8000	Dimmable, R30

Lamp data are from a major manufacturer; the listing is selective. Other manufacturers' offerings and data may be different. Linear and compact lamps are typically available in multiple color choices.
[1] Initial lumens for lamp only. Linear fluorescent at 85 CRI, compact at 82 CRI
[2] Efficacy when operating on typical ballast
[3] Rating at three hours per start

LED SOURCES

As a source of useful white light, LEDs are less than a decade old, although the roots of the technology extend back to the middle of the twentieth century. But, in a brief period of time, LED technology has proved an energy-efficient, versatile, and effective source of light. LED lighting offers the benefits of a small, well-controlled source of light, together with energy efficiency and long life (which combine for low operating cost).

LED Technology

A light-emitting diode is a semiconductor—a wafer of materials that supports electrical current in only one direction. The diode is "doped" (that is the technical term) so that one side has a negative charge and the other has a positive charge, with a junction, or active zone, in between.

How an LED Creates Light
When electrical current is applied, electrons flow from the negative to a positive charge. In the junction area, negatively charged electrons combine with positively charged "holes" and release energy, some of which is light (depending on the materials used) and some of which is heat.

TABLE 11.8 Summary of Attributes of Fluorescent Sources

Light Output	Wide range: 1400–5000 lm for linear lamps 450–3200 lm for compact lamps Excellent lumen maintenance
Luminous Efficacy	High: 95–100 lm/W for linear lamps 40–75 lm/W for compact lamps
Color	Many options: 2700–6500 K at 80+ CRI
Life	Long: 20,000–30,000 hours for linear lamps 6000–20,000 for compact lamps
Auxiliaries	Ballasts required for all lamps but are integrated in some compact lamps
Types	Linear, compact; compact screw base available on many incandescent types
Operation	Instant on; some warm up to full brightness; dimmable version dims with dimming ballast and compatible control

The released light is not white; instead, it is constrained in a narrow spectrum, depending on the materials used. Thus, an LED might emit red light (a red LED), or amber, green, blue, and several shades in between.

To get to white light, you need another step, and there are two basic methods:

1. Mix red, green, and blue LEDs, using the primary colors of light to mix to white. This method is subject to shifting color over time.
2. Convert energy from a blue LED by using phosphors that fill in the broad spectrum needed for white light.

Phosphor conversion predominates today due to its cost and consistency.

LEDs can last a long time (more on life later) and perform well in cold environments but are vulnerable to heat (either internally from energy that is not converted to light or from the external environment).

Drivers

LEDs generally need to be driven at direct current at low voltage rather than the alternating current at 120V in residential lighting circuits. An electronic power supply, called a *driver*, provides the proper adjustment to the electrical input. Like ballasts, drivers must be compatible with the LEDs they feed.

LED Product Configuration

LEDs grow up into complete lighting devices in stages (called *levels* by LED specialists). Since LED product literature often uses these terms, it is useful to know where they fit.

- The chip, or *die*, is the basic semiconductor.
- A *package* combines the die with fundamental thermal management, lead wires, and a protective lens that helps to extract the light from the junction or active zone in the die.
- An *array* combines the LED packages onto a *circuit board* ready for connection to the driver.
  - The *final product* assembles the array or circuit board together with the driver, additional optics to shape the light, and external thermal management to evacuate excess heat. Unlike other technologies, the final product here may be an LED lamp (designed for use in non-LED fixtures) or a dedicated LED luminaire.

LED Lamps

LED lamps—sometimes called *retrofit lamps*—are intended to replace conventional incandescent or fluorescent lamps in conventional fixtures. Typically, the LED lamp is designed to replicate the performance of a conventional light source with the benefits of higher efficiency, longer life, or other operational advantages.

For incandescent replacement, LED lamps incorporate a driver in the base of the lamp, enabling operation from the incandescent lamp holder. LED replacements for low-voltage lamps use a driver that operates off the low-voltage transformer.

Replacements for fluorescent lamps may incorporate a driver that operates off the fluorescent ballast, bypasses the ballast to receive line-voltage supply directly, or installs separately (replacing the ballast).

LED Luminaires

Dedicated LED luminaires use the LED array or circuit board as the light source, usually with a separate driver housed in the luminaire. The LED luminaire may replicate a conventional luminaire (e.g., a recessed downlight), or it may take advantage of LED technology to create form and function altogether.

Character of the Light

LED lamps and luminaires begin with very small sources of light, which need to be aggregated in order to produce the quantity of light needed for various lighting tasks.

Clustered in a small array and shaped by precise optics, LED sources can produce well-controlled beams of light typical of halogen PAR and MR lamps. Using diffusing optics, LED light can be spread in a distribution similar to general service incandescent lamps or soft-edged downlights.

Most tubular LED lamps emit light from an arc of 150 to 180 degrees, whereas linear fluorescent lamps emit light from 360 degrees. LED arrays in large luminaires easily replicate the distribution of fluorescent versions.

Glare

Any light source can cause unpleasant glare if it creates excessive brightness in the field of view, especially if seen against a much darker background. LED sources, particularly directional types, are more prone to glare than most other types because of the small physical size of the chip and the ability to concentrate its brightness. Looking into a poorly shielded beam of LED light is quite uncomfortable.

Color

LED sources can achieve virtually any tone of white, from warm 2700 K to very cool 6500 K. The color depends largely on the phosphor formulation. To achieve a warm tone, more of the blue LED light needs to be converted by phosphor, which uses a little more energy than is required to create a cool tone. Thus, warm LEDs produce a little less light (and lower efficacy) than cooler LEDs.

For residential applications, LED products typically are offered in 2700 K and 3000 K. Warmer tones (2200–2500 K) are available in some specialized products. So-called neutral 3500 K and cooler 4000 K, which are more appropriate for commercial applications, are also available.

Achieving good color rendering depends on the phosphor mix. There is a large selection of LED products offering at least 80 CRI; the choice of products with 90 CRI today is both more limited and generally more costly.

Experience with early generations of LED products, particularly those of poor quality, raised concerns about large *inconsistencies* in color. This issue is discussed in more detail under "Product Quality." For now, understand that LED technology actually can improve color consistency, but the result depends on how well the product is designed and manufactured.

LED sources also can create colored light, using LED chips without phosphor coating. A luminaire with controllable red, green, and blue LEDs can create a wide range of colors and change them on command. The effects can be luxuriously subtle or dramatically inspiring. (see Figure 11.9). We look at this in more detail in Chapters 12, "Light Fixtures," and Chapter 13, "Lighting Controls."

Light Output

In the last five years, LED technology has improved to the point that LED lamps and luminaires can deliver the lumen output needed for typical residential lighting requirements. The relatively low light levels needed for most residential applications (compared to commercial or industrial use) also make LED lamps an option for residential lighting.

High-output LED sources face two challenges: the cost of the LEDs themselves and managing the heat from high-wattage operation. Here the small size of the LED limits its ability to exhaust heat and therefore constrains its current and wattage. Small incandescent sources, such as a decorative or MR16 lamp, often provide higher lumen output than LED replacements. LEDs are far more efficient, of course. See Table 11.9 comparing the two light sources in both residential grade and high output varieties of LED.

With a few exceptions, LEDs provide better beam control and uniformity than conventional sources.

FIGURE 11.9 Colored LED lighting in a bathroom
Courtesy of Kichler

TABLE 11.9 Comparing Incandescent and LED

	Incandescent	**LED**
Type	MR16	MR16
Wattage	50	10
Light output	800 lm	500 lm
Life	3000 hours	25,000 hours
Type	Decorative (Hal)	Decorative
Wattage	40	3.5
Light Output	540 lm	1800 lm
Life	2200 hours	25,000 hours

Life

Since they lack a vulnerable part, like a filament or cathode, LEDs can emit light for a long time. LED *chips* can operate for 100,000 hours or more. Although the LED continues to operate, however, the light output depreciates; ultimately, the LED is no longer useful as a light source.

Useful Life

To address this, the lighting industry has changed the definition of rated average life for LED products from hours to failure (conventional sources) to hours to end of useful light (LED).

The accepted standard for LED sources is 70 percent of initial lumen output. This is called L_{70}. Note that this is a consensus value for rating life; users might consider higher (L_{90}) or lower levels (L_{50}) as more appropriate for determining useful light in specific applications.

The L_{70} value means the number of hours of operation until the light output of the LED chip depreciates to 70 percent of its initial lumen rating (i.e., until the LED loses 30 percent of its light output).

Some LED lamp and luminaire manufacturers publish L_{70} chip ratings as if they represented the life of their *products*. Since the LED product combines an LED array, electrical connections, driver, optics, and other components, looking at the chip alone does not give a true picture of product life.

Instead, manufacturers can combine the lumen depreciation of the chips with life expectancies of all other components to arrive at a combined life rating. At this point, it makes sense to remind users that these ratings are averages that reflect the inherent variation in manufacturing process.

This rating translates into the hours at which half of a large sample of products is delivering at least 70 percent of initial light output and half is not. Of those that are not delivering 70 percent of light, some will be dim due to lumen depreciation and some will be dark due to component failure. The actual situation is more complicated, but this explanation is more realistic.

Measuring Light versus Conventional Sources

Photometry of LED lighting differs from that of conventional sources in three important ways and is governed by two important standards: LM-79 and LM-80, created by the Illuminating Engineering Society of North America.

1. Lumen depreciation for an LED chip is *estimated* based on a limited number of measurements (taken over a minimum of 6000 hours). After the test period, the rate of depreciation is *extrapolated* according to a standard formula to the L_{70} point.
2. Light output and life ratings need to be correlated to the temperature the LED chip experiences inside the lamp or luminaire.
3. Light output for luminaires is measured with *each* specific LED chip used. Conventional photometry is measured relative to a standard lamp (of the appropriate type) can be adjusted for different lamp ratings.

The key point: You need to look carefully when comparing LED and conventional products because they are measured differently.

Impact of Temperature

The temperature in and around LED products significantly affects both the light output of LED chips and the failure rate of electronic components generally. As internal temperature rises, LED light output diminishes. For this reason, lumen depreciation of the chip is measured at various temperatures, and the chip results are correlated to the temperatures that the chip will experience when assembled into a lamp or luminaire.

Dimming

LED chips dim easily: A special dimming driver either reduces the current flowing through the LED or adjusts the voltage. We discuss the details of dimming control of LEDs in depth in Chapter 13 "Lighting Controls," but here is a quick look at three critical issues.

1. Unlike incandescent lights, standard LEDs do not create a warmer tone of light when dimmed. For task applications, this is probably not a problem. But for creating a sociable mood and atmosphere, it is a drawback. There are, however, LED products that do warm up as they dim (discussed below).
2. The LED lamp or luminaire must include a dimmable driver that is compatible with the LEDs it operates (similar to the requirement for dimmable ballasts in fluorescent lighting). Look for LED lamps that are specifically rated for dimming, and specify dimmable drivers in LED luminaires.
3. The driver and dimmer must be compatible. With the multitude of dimmers available and the rapid changes in driver technology, this can be a problem. Some incandescent dimmers will handle LED products effectively; some will not. Careful specification is required.

To emulate the warm, cozy mood associated with dimmed incandescent light, you can use specially formulated LED lamps and luminaires, whose color temperature lowers as they dim. These "warm dimming" products typically include both white and amber LEDs. The dimming signal dims the white ones more than the amber, resulting in a steadily warmer light.

It is also worth noting that when dimmed to the lowest setting, most LED products do not appear as dim as incandescent.

LED Options: Rapidly Changing Technology

New LED lamps and luminaires reach the market almost daily, offering more ways to light spaces, reduce energy and maintenance cost, and enhance lighting effects. LED sources continue to increase in light output, efficacy, color quality, life, and dimming capability, all at falling costs.

Most—but not all—incandescent lighting has LED counterparts. Some of these offer advantages of size, efficiency, control, and color options. Others fall short on one or more attributes.

Rapid change—including these dramatic improvements—brings with it confusion, frequent and speedy obsolescence, and ever higher expectations. To that end, you can look forward to further developments.

Table 11.10 is, at best, a snapshot of LED lamp options. LED luminaires are covered in Chapter 12, "Light Fixtures."

PRODUCT QUALITY

Incandescent light sources are more than 100 years old, fluorescent are more than 60. By comparison, LED lighting is still young. The quality of lighting products reflects their age. In this discussion, the term "quality" refers to reliability and consistency with expectations, whether published ratings or unstated assumptions.

TABLE 11.10 LED Lamp Options

LED Lamp	Lumens	Efficacy (lumens per watt)	Life (hours)	Notes
7A19	470	68	25,000	20% more light than 29A19 (Hal)
11A19	830	75	25,000	10% more light than 43A19 (Hal)
19A19	1780	94	25,000	19% more light than 72A19 (Hal)
3.5B11	180	51	25,000	Clear, candelabra
8R20	530	67	25,000	39% more light than 40R20 (Inc)
13BR30	750	9.5	25,000	21% more light than 65BR30 (Inc)
8PAR20	470	59	45,000	Lower lumens, comparable intensity
12PAR30	850	71	45,000	Equal light
19 PAR38	1200	17.8	45,000	10% more light
5.5MR16	290	58	25,000	LED MR16 lamps perform comparably to halogen MR16s of roughly three to four times the wattage, but you should evaluate visually
7MR16	380	54	40,000	
10MR16	485	49	25,000	
17T8	1600	94	40,000	4' lamp

Lamp data are from a major manufacturer; the listing is selective. Other manufacturers' offerings and data may be different. A and BR lamps listed are 2700 K; PAR, MR lamps listed are 3000 K. Efficacy listed includes driver.

Over time, users and producers—the market—have sorted out acceptable quality, performance, price, and value resulting in a range of *identifiable* options, following the maxim, "You get what you pay for."

LED products have not yet had enough market exposure to sort out the good from the bad or, perhaps, just the ugly. For lighting products generally, and LED products specifically, users face two risks in terms meeting expectations: color and life.

Quality of Color

Current standards allow for a great deal of variation among light sources nominally of the same color temperature. Two sources rated with the same color temperature need not appear the same. Indeed, often they do not. Moreover, each manufacturer applies its own tolerances (tight or loose), so even products from a single manufacturer may appear different—a particularly galling problem when they are viewed next to each other.

Visible variation has been most challenging for LED sources, although CFL lamps may also fare poorly in this regard. Manufacturers of linear fluorescent lamps have narrowed the tolerance for their lamps successfully, although each color point is different.

Evaluating products visually and using reputable brands best protect against surprise and disappointment. Energy Star–qualified products and items carrying the Lighting Facts label require third-party verification of the information provided. They too can be a viable indicator of quality.

Although the color of fluorescent light remains fairly stable over the long life of the lamp, LED sources may experience notable color shift over their lives. Elevated temperature around the LED appears to be the most likely cause of the color change. (Heat is a problem for LEDs in many ways.)

Quality of Life

How does lamp life compare to expectations? Do you expect half of your lamps not to reach their rated life? That, in reality, is the definition of rated life.

Incandescent and fluorescent lamps—even those with current improvements—have enough history that their life ratings are generally reliable, perhaps even conservative in the case of linear fluorescent lamps. However, frequent switching and temperature fluctuations in the case of integral CFLs can lead to shorter than expected life.

The life of LED products is estimated based on extrapolated measurements of lumen depreciation, correlated to the temperatures experienced in the lamp or luminaire. In some cases, the expected failure rate of electronic and other components factors into the final life rating. Clearly, there is a lot of room to fudge or misestimate the life rating. Once again, brand may be a necessary stand-in for reliability.

Exercise: Light Sources

This exercise involves observation of specific light sources and their effects. Describe what you see (not what the chapter tells you should see). You can use your own home or any other space (ideally a kitchen or bath). Use your journal to record your observations.

Describe the function of the lighting, the quality of light (including brightness, distribution, and color), and the luminaire for each type of lighting listed:

1. Incandescent general service lighting
2. Incandescent directional lighting
3. Compact fluorescent lighting
4. Linear fluorescent lighting
5. LED lighting

SUMMARY

The three electric light sources appropriate for kitchen and bath spaces are incandescent, fluorescent, and LED (light-emitting diode). Once you are aware of the features and benefits as well as the limitations of each source, you can choose the best option(s) for your lit spaces. Lighting is a key element of design. As lighting technologies continue to evolve, the only limits will be your imagination.

REVIEW QUESTIONS

1. What are the key strengths and weaknesses of incandescent light sources? (See "Attributes of Incandescent Light Sources" pages 148–149)
2. What are the benefits of halogen and infrared coated halogen technologies? (See "Halogen Technology" pages 153–154)
3. What are the qualities of fluorescent light sources? (See "Fluorescent Lamp Options" pages 163–164)
4. What are the benefits of using LED light sources? (See "LED Sources," and "LED Options: Rapidly Changing Technology" pages 164 and 169)

Light Fixtures

Chapter 11, "Comparing Electric Light Sources," gave you information on what light source to choose. The features and benefits of these light sources have to interface with the style and function of the fixture you decide to specify. The term "luminaire" often is used interchangeably with "fixture" in the lighting industry, as in the terms "recessed fixtures" and "recessed luminaires."

> *Learning Objective 1: Recognize what important factors will assist you in researching and choosing the correct fixture.*
>
> *Learning Objective 2: Evaluate fixtures based on functional lighting requirements of the project.*
>
> *Learning Objective 3: Identify appropriate locations for fixtures based on functional and aesthetic requirements of your lighting design.*

LIGHT FIXTURE SELECTION CRITERIA

Once you have your spaces planned out with regard to furniture and fixtures, you need to decide how you will approach the lighting aspect of the project. Whether it is new construction or an existing space that is being renovated, you should consider the next 10 items before you start your selection of fixtures for your project. Some of the items are based on a functional approach; others are based on aesthetics.

1. Existing electrical
2. Quantity of light
3. Fixture lamping
4. Materials and finishes of the space
5. Fixture finishes
6. Style
7. Budget
8. Timeline
9. Ceiling height
10. Size and proportion of fixture

Existing Electrical

Renovating a kitchen can impact the existing electrical. Going from a couple of centrally located fixtures to a ceiling filled with recessed lighting along with under-cabinet lighting and new appliances is a big change and can create a larger demand for electrical service (see Figure 12.1). Depending on the age of the home, this may not be possible. In that case, the existing electrical service may need to be upgraded. This can impact the budget significantly.

Quantity of Light

The quantity of light is referred to as the lumen output. Determining the functional and aesthetic requirements of the space will help you determine what lighting levels you need.

Fixture Lamping

Once you have decided on how much light you need in your space, you must choose a fixture that will give you this light. A surface-mounted fixture with three or more open sockets for lamping will give you more light than a fixture with one or two open sockets (see Figure 12.2).

Materials and Finishes of the Space

The materials and finishes used in the space impact the end result of the lighting. If the space uses light colors in its composition, light will bounce around more than one with dark colors, which will absorb the light. In addition to color, texture also impacts light. Smooth shiny materials, such as the desktop seen in Figure 12.3, reflect more light than textured matte surfaces, although glare might be an issue. More light would be required in a space that is dark and textured. To achieve the same brightness in the room, charcoal grass cloth with a dark floor would need more light than a lighter-value wall color in a semigloss. Another example of materials with contrasting reflectivity would be a honed black slate compared to a polished Carrara marble.

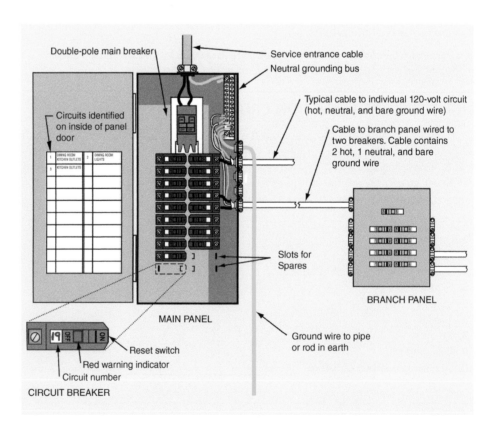

FIGURE 12.1 Electrical panel

FIGURE 12.2 The multiple socket light fixture above the mirror provides light for grooming and also contributes to light in the shower area.
Design by Terence Tung, codesigner Joseph Irons, CGR, GMB, CAPS, CGP, Kitchen Craft Cabinetry, Newcastle, WA
Photo by Tom Redner

FIGURE 12.3 The smooth, shiny surface on the desk in this space reflects more light than a matte surface would.

Design by Tammy Mackay, AKBD, Design Eye Ltd., Edmonton, AB

Light reflectance value (LRV) is how much light is reflected off a surface. Usually LRV is expressed as a percentage. Paint companies include this information with their color offerings. For example, a surface painted Benjamin Moore Ashley Gray reflects about 33 percent of the light hitting it where a surface in Benjamin Moore Linen White reflects about 83 percent.

Fixture Finishes

The finishes of the fixtures are often chosen because of prior finish selections. For example, if oil-rubbed bronze was chosen for the plumbing faucets and hardware throughout the space, then oil-rubbed bronze may be the finish of choice for the frame of the light fixture. It is not always the case; however, repeating finishes enables you to achieve a harmonious, successful space. Knowing where the line is drawn between the monotony of sameness and variety creating interest requires the lighting designer to be thoughtful in his or her choices.

Shade finishes are an important consideration and can affect the amount of light output. Figures 12.4 and 12.5 show similar fixtures, but because one shade is translucent and the other is more opaque, the amount of light emitted from the fixture is different. Some shades are completely opaque with the light only radiating from the opening at the bottom of the shade.

Style

During the initial part of the design process, information is gathered to create a complete program for the project. This program is the basis on which decisions will be made. The lighting package should reflect the style and atmosphere of the space. When choosing suitable fixtures, consider materials and finishes already selected for the project to help you make decisions. Whether the space is modern (see Figures 12.6) or traditional (see Figure 12.7), a number of manufacturers carry a style that will interface with your design intent. Layering different kinds of fixtures in one room requires an eye for design. Consulting a professional can be a valuable experience.

FIGURE 12.4 Translucent pendant fixture

FIGURE 12.5 Opaque pendant fixture

Budget

Once you are at the stage of choosing and purchasing lighting, the budget has already been set. Being creative with your budget is essential to a successful lighting project. There are great choices at every price point. If the budget does not allow the desired pieces, a less expensive alternative to the decorative luminaire may be purchased and changed out later when budget permits.

FIGURE 12.6 Modern chandelier fixtures
Courtesy of Kichler

For comparison purposes, let's use a scale out of 10 as an indication of budget. 10 is a high lighting budget, 5 is mid-range, and 1 is low. If your budget for lighting is a 7/10, rather than choosing all 7/10 fixtures, you can choose a 9/10 decorative fixture for an important focal point and select some 6/10 ones for less important areas. Regardless of the approach, having a plan in the initial stages of the project is important. Fixtures are installed in the later stages of construction when budgets are stretched and people's patience is put to the test. Ensuring lighting is planned and budgeted for in the beginning will help eliminate disappointment later.

Timeline

Depending on where fixtures are manufactured, they can take several weeks or even months to arrive. The lighting order should be placed immediately after the project is started to avoid any delay in the final installation of the fixtures. There may be off-the-shelf options available that meet functional considerations, but a readily available alternate that has the same style and finish as the desired fixture can be very difficult to find. If the timeline of your project is short, choosing from a company that has a large stock of items or can get all the lighting components quickly may be your only option.

FIGURE 12.7 Traditional chandelier fixture
Courtesy of Kichler

Ceiling Height

Knowing the ceiling height of each room you are choosing lighting for is vital. A fixture you pick for an 8-foot-high ceiling is completely different from one you would pick for an 18-foot-high ceiling. Some recessed fixtures may simply not have enough lumen output to get light from way up high down to a more human height level. For the higher ceiling, you may need fixtures with more light output, not more fixtures. Keep in mind also that changing the lamping at a later date is possible. For high ceilings in foyers, you may want to add the option for the fixture to be raised and lowered for cleaning and ongoing maintenance.

Size and Proportion of Fixture

Proportion and scale are important design principles to keep in mind when choosing your fixtures. A towering ceiling height of 18 feet requires a larger ceiling fixture that is in proportion to the space. A 3-foot-tall fixture would look out of place and insignificant. When light fixtures are hung over furniture, size and proportion are also essential. For example, a dining table should have a fixture that is in proportion to its size. A piece that is too large looks odd and top heavy. A piece that is too small looks weak. The right solution may be hanging two or three smaller fixtures in a row. This information is crucial early on in the project as more the electrician will have to install more than one junction box. A scaled sketch elevation of the room or table along with a form showing the general shape and size of the fixture may be all you need to visualize whether it is a suitable choice (see Figure 12.8).

FIGURE 12.8 A scaled sketch can help the designer to visualize how a fixture will look in the space. *Design by Tammy MacKay, AKBD, Design Eye Ltd., Edmonton, AB*

TYPES OF FIXTURES (LUMINAIRES)

There are different types of luminaires available. The type you specify is dependent on the installation and application in addition to aesthetics. They are categorized by their mounting methods and are:

- Recessed
- Ceiling or surface mount
- Semi-surface mount
- Suspended
- Wall mount
- Portable

Recessed Fixtures

The category of recessed lighting captures all fixtures that are recessed into a cavity of space. This cavity can be in the ceiling, as seen in Figure 12.9; in a wall, floor, or stair assembly (see Figure 12.10); in a niche or in millwork. Typically, the fixture contains a housing component that gets hidden in the cavity and a trim piece that attaches to the ceiling. The trim of the fixture is the component you see on the surface. The fixture is hardwired into the electrical system. If your project is new construction, your options for using recessed fixtures are limitless. If it is a renovation project where ceilings are finished, you have to make sure you find a fixture that will allow you to recess it into the ceiling space without disturbing the existing ceiling and finishes. In addition to this, if the space in which you are recessing the housing is insulated, you must use a type of fixture can be recessed into an insulated space. Using an incorrect fixture can be a fire hazard.

FIGURE 12.9 Two of the recessed lights in this bathroom are directed at the tile wall to highlight the iridescent quality of the tile.

Design by Janice Stone Thomas, ASID, CKD, codesigner Alia Richards, Stone Wood Design, Inc., Sacramento, CA Photo by Dave Adams Photography

Recessed downlights are a great choice for creating lighting effects like spotlighting, grazing, and wall washing. Spotlighting typically is used to highlight artwork. Grazing is used to highlight a wall of texture like a stacked stone. Wall washing washes out the texture and creates a lit wall that is smooth and even. It makes the room feel spacious and draws attention to architectural details.

Recessed lighting typically is functional rather than decorative. What you are lighting is what grabs your attention, not the fixture itself. There are exceptions, however, such as the fixture seen in Figure 12.11, which has a basic housing but has a crystal trim that gives a subtle sparkle on the ceiling.

FIGURE 12.10 Recessed stair fixture
Courtesy of Eureka Lighting

FIGURE 12.11 Recessed fixture with trim
Courtesy of Eureka Lighting

Matching the recessed lighting trims to the surface into which it is being recessed is the best choice. Manufacturers offer many choices to make this easy. The majority of recessed fixtures are used for general/ambient light and accent lighting.

Ceiling or Surface Mount

Surface-mounted fixtures typically are mounted on the surface of the ceiling. They require a junction box and are hardwired to the electrical system. A surface mount can be as simple as the large fixtures seen in Figure 12.12, or it can be a rail system that has many components, like the one shown in Figure 12.13. The rail system can be your only overhead lighting choice if you have vaulted ceilings with no space to recess a fixture or junction box. Surface-mount rope lighting or tape lighting (see Figure 12.14) is used for lighting details, such as coves, where it is mounted to the surface of a concealed area. Most surface-mount fixtures are used for general/ambient light.

FIGURE 12.12 Ceiling mount fixtures
Courtesy of Kichler Lighting

Semi-Surface Mount

A semi-surface-mounted fixture as seen in Figure 12.15 looks much like a surface mount, but it has an extension on it so that it drops farther from the ceiling. This type of fixture often is used when ceiling heights are 8 feet and the fixture will be placed where people are walking underneath it. A decorative fixture is desired, but there is not enough headroom clearance for a chandelier. An example would be using it in a single-story front entrance. It brings a little more attention to the foyer. A matching surface mount might be used in an adjacent hall.

Suspended Fixtures

Suspended fixtures can be one of the most striking features of your lighting design. They provide a huge impact in a space as well as sculptural interest and sparkle (as in the French country kitchen in Figure 12.16). Suspended fixtures are hardwired to a junction box located in the ceiling. Depending on the design and finish of the fixture, light can be directed in many different ways. (See Figure 12.17.) Shades can direct the light down (see Figure 12.18) or up. Some suspended fixtures, called direct/indirect fixtures, do both. You can choose what percentage you want up and down: for example, 80 percent up/20 percent down.

Placement Tip

Under-cabinet lighting should be placed closer to the front of the cabinet to ensure that the majority of the countertop is washed in light. LED tape lighting is a great option for this. Use a fascia piece or strip of wood to hide the light source from view.

FIGURE 12.13 A rail system is
effective in this space because the
ceiling is vaulted and includes multiple
skylights.

*Design by Tammy MacKay, AKBD,
Design Eye Ltd., Edmonton, AB*

Indirect lighting reduces glare. Reducing glare is becoming a more important component
of lighting design as we are using polished surfaces like quartz and granite more than
ever. Glare reduction is particularly important for the aging eye. We are also using technology in nearly every room of the home, and glare on displays can be uncomfortable to
the eye.

Suspended fixtures are categorized in four ways:

1. *Chandelier.* Larger than 18 inches in width or diameter with more than one socket for a lamp

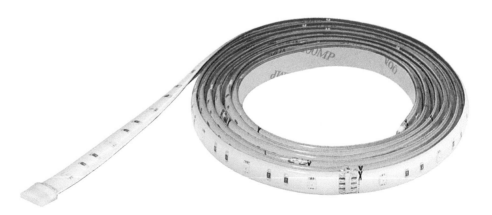

FIGURE 12.14 Linear LED tape
Courtesy of Hafele

FIGURE 12.15 Semi-surface mounted ceiling fixture
Courtesy of Kichler

FIGURE 12.16 The chandelier in this French country style kitchen is a beautiful component of the design.
Design by James E. Howard, CKD, CBD, Glen Alspaugh Co., LLP, St. Louis, MO
Photo by Alise O'Brien Photography

Placement Tip

Consider these four tips:

1. A general guideline is that the width of a chandelier should be offset inside the edge of the table by at least 12 inches on all sides.
2. Above a table: 30 to 36 inches for an 8-foot ceiling. Raise the chandelier 3 inches more for each foot higher than 8 feet.
3. Above a raised bar: 66 to 72 inches above the finished floor.
4. Beside the bed: Ensure the bottom of the shade sits at shoulder height when you are in a reading position in bed.

2. *Chandelette.* Eighteen inches or smaller in width or diameter with more than one socket for a lamp
3. *Pendant.* A suspended fixture with one socket for light source
4. *Pendalette.* A miniature version of a pendant

Chandeliers and pendants are not for just the dining room and foyer anymore. We are seeing them in bathrooms over vanities and tubs, over bedside tables, and over kitchen islands. Outdoor pendants and chandeliers are becoming more readily available as well. They are exciting to work with and provide function as well as an artistic expression to any space.

FIGURE 12.17 The more opaque pendants in this kitchen direct most of their light downward while the translucent fixtures above the island allow light to spread in all directions.

Design by Bryan Reiss, CMKBD, codesigner Susanna Caongor and Diane Murphy, Distinctive Design, Mt. Pleasant, SC
Photo by John D. Smoak III

FIGURE 12.18 The opaque shades on the chandeliers direct most of their light to the islands.
Design by Sandra L. Steiner-Houck, CKD, Steiner & Houck, Inc., Columbia, PA
Photo by Peter Leach

Wall-Mount Fixtures

Wall-mounted fixtures are great lighting choices. (See examples in Figures 12.19 and 12.20.) Depending on the design of the fixture, it can provide an indirect or a direct light source. In some situations, it may not be possible to install ceiling fixtures because of height restrictions.

FIGURE 12.19 Wall-mounted fixtures
Courtesy of Kichler

Placement Tip

Consider these two tips:

1. Ideally, wall-mounted fixtures over a bathroom vanity should be mounted on either side of the mirror, and the light source should be just above eye level. This location will help to eliminate any unflattering shadows on the face when applying makeup or shaving.
2. In commercial spaces, to be compliant with guidelines from the Americans with Disabilities Act, wall light fixtures projecting from walls with their leading edges between 27 and 80 inches above the finished floor shall protrude no more than 4 inches into halls and passageways.

Specific wall fixtures for picture lighting are a great choice as they can focus the lighting exactly where you want it, especially if you have soaring ceiling heights. They are excellent choices for general/ambient, task, and accent lighting. Wall-mounted fixtures also are used extensively in outdoor applications.

Portable Fixtures

Portable fixtures include table (Figure 12.21) and floor-type fixtures (Figures 12.22 and 12.23). They are often overlooked during the lighting design phase as they do not require any prewiring. Layering these fixtures into your lighting design can add to the atmosphere

FIGURE 12.20 Wall-mounted fixtures
Courtesy of Kichler

FIGURE 12.21 Table fixture
Courtesy of Kichler

FIGURE 12.22 Traditional-style floor fixture
Courtesy of Kichler

FIGURE 12.23 Modern-style floor fixture
Design by Christine Pandur, AKBD, and Tammy MacKay, AKBD, Design Eye, Ltd., Edmonton, AB

and drama of the space. Shades for these fixtures can be a variety of materials, from translucent linen to an opaque material. Translucent shades provide more of a soft glow whereas opaque shades direct the light. The ability to move these lights around makes them very versatile choices. Portable fixtures are used for general/ambient lighting as well as task lighting.

Comparing fixtures

Visit a local lighting showroom and find fixtures by type:

1. Find surface-mount fixtures that look the same but have different light output because of the number of lamps required for the fixture. Explain why you would select one or the other for a particular design application.
2. Compare different pendant shade finishes. Some shades are opaque, and some are translucent. Explain how this affects the amount of light output of the fixture.
3. Note how the showroom is laid out. How are the fixtures grouped?

   By type. For example, all suspended fixtures together and all surface mount, and so on

   By finish. For example, all oil-rubbed bronze fixtures together and all satin and nickel finishes, and so on

   By manufacturer. Grouped by manufacturer so all finishes and types are together

SUMMARY

Choosing fixtures is a balancing act between ensuring the correct amount of lighting is achieved and your aesthetic goals are met. The basic design elements and principles assist in making choices regarding the aesthetics of your fixtures, and the initial programming helps you to make functional choices. The number of lighting options is overwhelming. Getting it right in the beginning saves time and money in the end. Being armed with selection criteria can help you narrow it down to a lighting package that is not only beautiful but also meets the needs of your project. There is no perfect fixture that can do it all for you. A thoughtful layering of all the fixtures discussed so that you are combining downlighting with indirect lighting and focused lighting with general lighting gives your project the refined look that you are after. If there are safety concerns, always check with local authorities that have jurisdiction for regulations with regard to lighting and electrical installations.

REVIEW QUESTIONS

1. What are some items to consider when choosing fixtures? (See "Light Fixture Selection Criteria" page 173)
2. What is LRV, and why is it an important point to consider? (See "Materials and Finishes of the Space" pages 174–176)
3. Where ceilings are vaulted and there is not a lot of space to recess lighting, what type of lighting might be suggested? (See "Ceiling or Surface Mount" pages 182–186)

Lighting Controls

Daylight is dynamic; it changes daily and seasonally. Electric lighting is static, until you add controls. Controls enable you to vary lighting to suit the task or create the mood, to reduce energy and maintenance cost, or simply to make life more comfortable and convenient. For the best results, controls need to be considered at the very outset of the lighting design. The process begins by establishing objectives and developing a plan and winds up by implementing it with the appropriate equipment and arrangement.

> *Learning Objective 1: Apply the benefits of controls to kitchens and baths.*
>
> *Learning Objective 2: Communicate using technical controls vocabulary.*
>
> *Learning Objective 3: Explain how controls work.*
>
> *Learning Objective 4: Distinguish controls options and strategies as they apply to kitchens and baths.*

All lighting requires control. At the least, you turn every light in the space on or off by toggling a single switch up and down. That is pretty basic: The room is either flooded with every available footcandle, or it is totally dark.

More effectively, each *layer* of light is controlled by a dimmer, permitting you to balance the brightness among the layers, tuning the overall lighting effect for the various activities, moods, and people in the space.

BENEFITS AND OBJECTIVES

Controls—switches, dimmers, and systems of linked devices—offer four primary benefits. These resolve into the different approaches to controlling lighting in a kitchen or bath.

1. *Mood/atmosphere.* This refers to adjusting the amount (and sometimes the tone) of light to enhance the emotional experience provided by lighting in the room. Control over the different lights in an open plan kitchen enables you to set a mood for casual social or family interaction or for more formal entertainment.
2. *Task tuning.* This involves adjusting lighting to support different visual tasks. Conceptually, this objective is similar to adjusting for mood/atmosphere. In practice, however, it is more a matter of how much light is directed to a task.

3. *Economy of operation.* This involves controlling lights to reduce energy consumption or to extend lamp life.
4. *Convenience.* This refers to controlling lighting to save time and concerns associated with entering an unlit space. Linked devices located at the multiple entries of a kitchen permit you to turn lights on and off conveniently, regardless of how you move through the space. (This can be a simple three-way switch or a more sophisticated remote keypad.)

Rapidly Changing Field

The field of controls is changing rapidly, perhaps with even more impact than evolving LED (light-emitting diode) technology.

The dimming of theatrical lighting began in the first third of the twentieth century, more than 75 years ago. The earliest dimmers simply shunted some of the power away from the lighting, reducing light output but not the total amount of power consumed.

Modern control for residences began in the 1960s with the introduction of electronic circuits into wall box dimmers. By the 1990s, residentially scaled dimming systems made theatrical effects possible in the home. With the introduction of networked and wireless controls, dimming control became more powerful and convenient.

The dimming first of fluorescent and more recently of LED sources has added considerable technical challenges to delivery of consistently satisfactory results. Nevertheless, the combination of digital lighting (LED) and digital control (particularly wireless control) promises to make lighting more responsive to users than ever before.

This rapidly changing field often makes it difficult to incorporate products and techniques from different manufacturers into your design. To try to make it easier to design and specify controls systems, this chapter begins by reviewing concepts and terms using generic language.

Controls Concepts

Planning for lighting controls begins at schematic design. Consider how you are lighting the space, the layers of lighting.

Each layer of light should be regulated by its own control. Where a single layer—downlighting, for example—serves multiple tasks, multiple controls permit adjustment to serve each task.

Conceptually, you can think of the control of each layer in three dimensions:

1. What *effect* do you want to see in the light: on/off or variable?
2. What *input* do you want to the control: manual or automatic?
3. What *linkage* do you want among the controls: stand-alone or networked operation?

Effect
On/off controls offer just one lighted setting: on. Variable controls might be two settings (high and low), three settings (high, medium, and low), or continuously adjustable (typical dimmer). Variable controls offer more choices. On/off controls are simpler and less costly to purchase and install.

Input
Manual controls require that you touch the device each time you want to adjust it. "Touch here" might mean simply toggling the switch lever or perhaps pressing an application on your phone.

Actually touching each control individually can be cumbersome and time consuming. Collecting the touch points into a single device—a keypad—can solve this problem.

"Automatic controls" mean that lights respond to an external signal without your direct, manual intervention. Motion sensors and clocks are simple examples.

TABLE 13.1 Lighting Controls

		LIGHTING EFFECT	
		On/Off	**Variable**
INPUT	**Manual**	Switch	Dimmer
	Automatic	Motion sensor turns lights off when room is empty	Clock-activated dimmer gradually increases light for wakeup
	Hybrid	Garage door opener turns lights on	Audio-visual control signals dimming for viewing movie

Controls that respond to manual control of another system are a hybrid of manual and automatic. Perhaps you would like the lighting in your home to turn on when you press the garage door opener or the lighting to dim when you turn on your television. Table 13.1 summaries the types of lighting controls used today.

Linkage

A network is a system of linked elements that responds in a coordinated fashion. In a three-way switch, a third conductor connects two switches, enabling them to both toggle the lighting on and off. In more sophisticated controls, devices are connected by communication wires—or wirelessly—so that instructions sent from one device can be followed by several devices at different locations. In addition, a device—dimmer or switch—can be controlled from several locations throughout the home.

Linking controls through a network significantly enhances the convenience of using the lighting system and thus increases its functionality.

THE LANGUAGE OF CONTROLS

At this point, you have established objectives for the controls, identified the lighting layers to be controlled based on your lighting design, and determined the effects you want to achieve (on/off or variable) and how you will tell the controls what to do (manual or automatic, stand-alone or linked).

To discuss how to implement the controls plan with specific equipment, we first need to introduce some more specialized terms.

Load

The term *load* describes the lighting connected to a control. The load identifies the type of electrical device to be controlled (incandescent, fluorescent, LED light sources; fan or shade motors) and the *total* wattage connected to the control (five 20-watt [W] lamps or luminaires represents a 100 W load). Both switches and dimmers are governed by load maximums. Although a switch generally can control a variety of light sources (and other loads), a dimmer (or variable control) typically must be compatible with the load it is controlling.

This is a fundamental and critical principle: Controls must be compatible with the loads they control, both the type of load and its magnitude. You cannot assume that any control device will control any given load effectively. You need to verify their compatibility. The more sophisticated the control and system—think a networked control and LED luminaires—the more likely you are to encounter interference between incompatible electronics.

Channel or Zone

The terms *channel* (from theater and audio) and "zone" (from the heating, ventilation, and air-conditioning field) refer to a group of luminaires (or other loads) controlled *together*. "Channel" and "zone" are essentially interchangeable in basic usage.

Typically, each layer of lighting (downlights, pendants, under-cabinet task lights, etc.) should be a separate channel or zone, meaning that each layer is connected to its own control. This way, you can adjust each layer independently.

You can think of this grouped type of control as all of the downlights connected to a single dimmer or switch. In a large kitchen where different areas serve different activities, however, the downlights might be divided into *two* channels or zones, each connected to its own dimmer or switch.

The more channels or zones in your system, the more you can refine the control of the lighting. But clearly, this comes at some incremental expense for the equipment and its installation.

Power and Signal

All electric lighting equipment requires power, which is supplied over the electrical *circuits* in the home. An electrical circuit is a closed loop consisting of conductors for the electricity, loads that use the electricity, controls that switch or dim the loads, and a protective device (circuit breaker) to guard against overloading—and hence overheating the circuit.

Controls can be inserted within a circuit or, in the case of large installations (typically commercial), multiple circuits may be controlled as a single channel.

Circuit Facts

Circuits are sized by the protective device and measured in amperes (a 20-amp circuit). The installing electrician is responsible for ensuring that the conductors will not overheat before the protective device cuts off the power. That is, "undersized" conductors will not be protected by a higher-amperage protective device.

Residential lighting circuits consist of two conductors, the hot and the neutral, plus a ground. In many residential applications, the switch or dimmer is inserted only on the hot leg of the circuit, meaning the neutral is not connected through the control. A common way of expressing this is to say, "There is no neutral wire in the box" (meaning the outlet box into which the control was wired). As you will see, neutral connections are required for many advanced control devices.

Circuits are not the same as channels or zones, although the term may be (incorrectly) used that way.

The control operates on the electricity that reaches the load. (Sometimes this is clarified as a load controller.) But how do you tell the control what to do?

In a simple dimmer or switch, the manual adjustment (moving the toggle lever or slider) opens the circuit (shutting off the lights) or signals the dimmer to adjust to a different level. Here the "user interface" and the load control are incorporated into one device.

In more complex systems, the user interface and load control may be separate devices, with a signal between them. In some cases, the signal travels over the same electrical conductors that power the lighting equipment (a power line carrier system); in others, there are separate signal conductors or a wireless signal.

The difference between power and signal is particularly important when considering sophisticated control systems and dimming LED or fluorescent luminaires.

Remember that communication requires both a sender and a receiver. The sender and receiver must be compatible, and the signal between them must be clear and free from interference.

User Interface

As discussed, the user interface is the part of the system you *touch* or tell what you want the control to do. It is where you input your command—manual, automatic, or hybrid. The user interface can be the familiar toggle lever, slider bar, or push-button keypad. Sometimes the user interface is integrated with the load control (the switch in your house); sometimes it is a separate device, such as your mobile phone, connected by a signal to the control.

Controller

A load controller adjusts the electrical input to the load and, through this, adjusts the light output. This is the component that does the actual work of switching or dimming. For simple control of incandescent sources, the load control and user interface are packaged into a single device—a wall box switch or dimmer.

With an LED source, it is a bit more complicated. Here the LED *driver* is actually dimming the LED emitters in the luminaire. The dimmer on the wall—the user interface—is telling the driver what to do. The topology for fluorescent dimming is the same: The dimming ballast dims the lamp (by lowering the current); the wall dimmer tells the ballast what to do.

As controls technology becomes more sophisticated, understanding the difference and relationship between the user interface and the controller is especially important.

Switch

A switch is a simple on/off device. It works by opening and closing the electrical circuit to the controlled load. Two three-way switches (also called three-pole) connect for control from two locations using three conductors; three four-way switches connect for control from three locations, using—you guessed it—four conductors.

An electronic switch takes a signal from another device on the control system. Multiple electronic switches typically can connect using just one additional conductor (the signal wire), permitting multiple locations with simplified wiring.

A relay is a switch controlled by a low-voltage signal, which *relays* the command from the manual or automatic input to the switch itself. Relays are used to control commercial lighting and residentially for window blinds and in some occupancy-sensing devices.

Dimmer

Wall box dimmers are installed in one or more outlet boxes (the same type of boxes used to install receptacles). Other types of dimmers, often called dimmer modules, are installed in a cabinet or dimmer rack, typically located in a closet or near the electrical service panel. Only the user interface—typically a keypad—is mounted in the space itself.

Although this approach is limited to commercial applications and very large residences, it is worth understanding the difference between the user interface (keypad) and the load control (dimmer module).

The user interface configuration includes wall-mounted keypads with touch screens that offer the user simple preset scenes they can select. A load control (dimmer module) when connected to a network allows users total flexibility to control any load they decide to connect to it.

Preset

A preset is a command that you can activate with a simple action, such as pushing a button. The preset recalls the light level you set each time you turn the control on or press the preset button.

With a basic slide-to-off dimmer, you turn the lights on by raising the slider to the level you want. You turn the lights off by lowering the slider to the bottom. Each time you turn the lights off, you "lose" the dimmed setting you had when the lights were on.

With a preset slider dimmer, you turn the lights on by pressing a button, bar, or a toggle. The lights turn on to the level set by the slide control. You turn the lights off by pressing the button, bar, or toggle again, leaving the slider itself alone. Now the dimmer will recall that setting when you next turn it on. Of course, each time you adjust the slider, the preset level changes.

Scene

A lighting scene is a setting for light in a space, typically using multiple layers of lighting. Conceptually, distinct lighting scenes should support each different activity in a space. The more scenes you need, the more important it is to have a comprehensive system that allows for this.

The more sophisticated systems are whole home approaches where the integration of lighting, electronics, and even window shades can be controlled from a single wall mounted key pad or a hand held device. This control can be accessed wirelessly as well with the main program operating from your home computer.

Remote

A remote is a user interface that enables you to signal the load control from a separate location. Technically, most keypads are remote devices, but typically they are just called keypads. A *hand-held remote* uses an infrared (IR) beam or a radio signal to communicate.

HOW DIMMERS WORK

Dimming depends on the load—that is, the source—being controlled. From a dimming perspective, several different loads are found in homes:

- *Incandescent.* These may be familiar line voltage types or low-voltage types operating on different types of transformers.
- *LED.* These can be dedicated LED luminaires or LED lamps designed to fit into incandescent fixtures. In either case, special drivers are required for dimming.
- *Fluorescent.* These can be dedicated linear or compact fluorescent luminaires or screw-base, retrofit compact fluorescent (CFL) lamps. In either case, special ballasts are required for dimming.

Dimming Incandescent Loads

All incandescent lamps perform about the same as they dim.

- *Dimming range.* Incandescent lamps can dim smoothly to very low levels. A deeply dimmed filament appears as a barely glowing twig.
- *Color.* Incandescent light changes color, moving from white to a much warmer amber tone (from 2700 Kelvin [K] or 3000 K down to about 2200 K).
- *Lamp life.* Incandescent lamps last longer when dimmed. Expected life of 1000 hours (less than a year in typical use) might double if dimmed by about 10 percent and quadruple if dimmed by 25 to 30 percent.

- *Efficacy.* Light output falls much faster than power consumption. For example, dimming power by 50 percent results in a drop of more than 75 percent of light (as measured). That is, incandescent sources are less efficient when dimmed than when operated at full output.

Different Dimmers for Different Loads

Incandescent loads fall into three basic types:

1. *Incandescent, line voltage.* Arbitrary (A) lamps, bulge reflector (BR), parabolic aluminized reflector (PAR), and basic decorative lamps are the simplest loads to dim and can be controlled with inexpensive dimmers.
2. *Incandescent, low voltage with* magnetic *transformers.* This application typically is found in many multifaceted reflector (MR) lights such as MR16 recessed downlights and linear low-voltage luminaires, such as cove or cable systems using MR16 or other 12- or 24-volt lamps.
3. *Incandescent, low voltage with* electronic *transformers.* This application is similar to other low-voltage types but is the more current transformer technology and is almost universal in track lighting today.

Dimmers specific to magnetic low-voltage and electronic low-voltage must be used. Standard dimmers will not work.

Line Voltage

Incandescent dimmers use a high-speed electronic switch that turns lighting on and off 120 times per second (twice per cycle). The longer the lights remain off, the dimmer the lighting effect (see Figure 13.1).

This type of dimming is often called by different names:

- Incandescent
- Triac (after the type of electronic switch)
- Forward phase (after the relationship of the switch to the electrical cycle)

These are simply different names for the same type of control.

Low-Voltage Magnetic

Dimmers for low-voltage incandescent loads with magnetic transformers differ from the simple incandescent type because they need to cope with the current induced by the magnetic components in the transformers. They can handle ordinary line voltage loads as well and are a type of forward-phase control dimmer.

This type of control is typically called a magnetic low-voltage (MLV) dimmer. Because a magnetic transformer is an inductive load (the magnetic *induces* a current), these devices may also be called inductive dimmers. MLV and inductive dimmers often require a neutral connection, which ensures proper operation (see Figure 13.2).

Low-Voltage Electronic

The inrush of current each time a forward-phase dimmer turns on causes problems for electronic transformers. As a result, electronic low-voltage transformers should be dimmed with reverse-phase dimmers, which allow the current to ramp up each half cycle (see Figure 13.3).

Forward Phase-Control

FIGURE 13.1 Forward-phase cut dimming

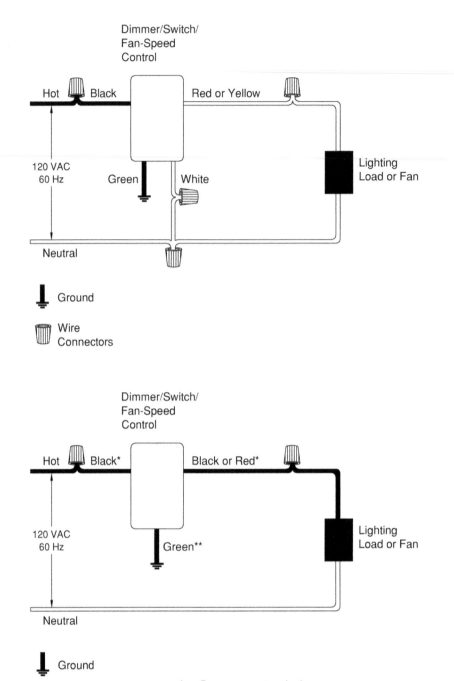

FIGURE 13.2 Diagrams of dimmers
with and without neutral wires

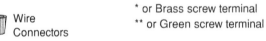

Reverse Phase-Control

FIGURE 13.3 Reverse phase cut
dimming

A reverse phase dimmer is also called an electronic low-voltage (ELV) dimmer and generally requires a neutral connection.

Dimming LED Loads

As you might expect with a fast-evolving technology, the dimming of LED luminaires and lamps does not match the experience of dimmed incandescent or, by and large, the consistency of fluorescent.

- *Dimming range.* High-quality LED luminaires can dim down to the 1 percent range needed for the best results. Other products—and LED lamps—do not dim as well.
- *Color.* Standard white LEDs do not change color as they dim. This is an important difference between LED and incandescent technology and is especially challenging where LED lighting is being used in place of incandescent. However, some LED lamps and luminaires are available with LED arrays that contain both white and amber or red LEDs. These arrays can dim to warm color. Such products may be valuable if you want to create the warm, glowing feeling associated with dimmed incandescent lighting.
- *Lamp life.* Dimming generally allows LEDs to operate at a lower internal temperature, which has a favorable effect on life. However, if product life is limited by *driver* electronics, the impact on life may not be significant. (They still will last longer than incandescent lamps, of course.)
- *Efficacy.* LED luminaires and lamps remain highly efficient when dimmed (actually increasing slightly due to lower internal temperatures).
- *Dimming issues.* Smooth dimming over a full range can be a challenge for LED lamps. Compatibility between the specific lamp and specific dimmer is critical, as is managing the load between minimum and maximum levels.

Dimming Drivers

Special dimming drivers dim LED lamps and luminaires. The dimming driver adjusts current and voltage to lower the output of the LEDs. Dimmers signal the dimming driver and tell it how to adjust the light output. Compatible dimmers are determined by the specific design of the dimming driver.

Dimming LED Luminaires

LED luminaires typically use dimming drivers either controlled over the power line by extra-low voltage (ELV) incandescent dimmers or controlled by separate signal wires. (These may be the analog 0-10V signal or a digital signal such as DMX 512, a theatrical protocol often used for colored lighting.)

Dimming LED Lamps

LED lamps with dimmable drivers rely on power line dimming control, either ordinary incandescent (forward phase) or ELV (reverse phase). Assuring *compatibility* between the specific LED lamp and the dimmer type, model, and manufacturer is critical to achieving satisfactory performance. To reduce the risk of problems, consult the compatibility charts published by both lamp and dimmer manufacturers and observe their load limits.

Dimming Fluorescent Loads

Fluorescent lamps vary more than incandescent lamps in terms of dimming quality.

- *Dimming range.* High-quality dimming systems can dim linear fluorescent lamps down to about 1 percent of light output. While these low levels are needed to create the perception of a dimmed environment, they carry a higher material cost to ensure consistent lighted results. Everyday dimming results in about 10 to 30 percent of light at the lowest setting. CFL lamps do not dim to as low levels as linear lamps. CFL lamps with integral ballasts have dimming issues of their own.

- *Color.* Fluorescent lamps do not warm in color as they dim, which is a very important difference from the dimming performance of incandescent sources. High-quality lamps are considered pretty stable in their color (although the change in the quantity of light can affect how we feel about the color).
- *Lamp life.* Fluorescent lamps do not enjoy the extended life of dimmed incandescent lamps but generally maintain their rated average life. Improperly designed dimming ballasts, however, can shorten life.
- *Efficacy.* Fluorescent lamps remain highly efficient when dimmed.

The dimming of fluorescent sources is performed by a special *dimming ballast*. The dimming ballast has a unique dual function. It *lowers* the current through the electric arc to dim reduce light output; at the same time, it *maintains* a steady current to heat the cathodes for proper operation and lamp life.

Fluorescent dimmers signal the dimming ballast how to dim. In residential applications, the signal often is sent over the power line, using either two or three conductors. In commercial applications, the signal often travels over dedicated control wires (either an analog 0-10V or a digital signal), which tends to be more reliable, although the separate wiring adds cost.

Choosing Ballasts and Dimmers

For pin-based fluorescent lamps, linear or compact, you specify the dimming ballast, which is supplied with the fixture, and the control. Fluorescent dimmer and dimming ballast *must be compatible*.

For retrofit CFL applications, you specify a dimmable lamp. Most of these are designed to dim on standard incandescent, forward-phase dimmers, although better results may be obtained with special CFL dimmers.

There is a multitude of dimmer styles, offering a variety of features from basic to sophisticated.

Sliders, Knobs, Toggles, and Rockers

You can set the dimmed light level using a variety of devices. Although they all perform the same function, they differ in style (contemporary to traditional) and ergonomic ease of use.

All dimmers today use electronic circuits. Most use an analog control; the farther you move the control, the more you dim. Digital dimmers convert taps or pressing into discrete light levels.

- *Slider dimmers* are easy to grasp and move, and their location on the dimmer back plate suggests the setting.
- *Rotary dimmers* provide a knob to set the light level. Some older clients may find that gripping the knob is not as easy as moving a slider, and it is difficult to read the setting from the position of the knob.
- *Toggle dimmers* look like toggle switches, which is their chief attraction. Due to the small size of the toggle, these types are not recommended for older clients.
- *Rockers* (sometimes called *paddles*) usually indicate that the setting is digital. Different taps, or how long you press the rocker, tell the dimmer how to set the light level. Rocker dimmers also match the appearance of rocker switches. Rockers are easy to touch, but their operation is not obvious to many people, making them hard to use in practice.

Preset Dimmers

As discussed earlier, a preset dimmer turns on to a previously established setting. This is a very convenient feature, useful in most applications. Preset sliders have a separate switch. Rotary dimmers typically push in for on/off control. Digital dimmers have a set button to create the preset level.

Smart Dimmers

Smart dimmers are dimmers that can retain multiple settings and respond to networked keypads as part of a system (see Figure 13.4). Smart dimmers offer a simple and inexpensive approach to system control.

FIGURE 13.4 Smart dimmer
Courtesy of Lutron

Illuminated Dimmers

How do you find a dimmer (or switch) in the dark, especially if you unfamiliar with the room?

Some dimmers offer a lighted model that glows when the dimmer is set to off. Digital dimmers often use LED indicator lights to show the current and preset levels. (The rocker does not tell you, of course.) The indicator lights also help you locate the control.

Lighted dimmers can either be friendly convenience or an annoying distraction in the dark. Find out your client's preference.

SENSORS

Sensors detect activity in the environment and signal lighting to respond by turning on, turning off, or changing output.

The two common types of lighting sensors include:

1. Motion sensors, to turn lights on or off, based detecting people in the space
2. Photo sensors, to switch or dim lighting, based on the amount of daylight detected by the sensor

Sensors have seen little application in residential kitchens and baths, but they play an increasingly important role in commercial control strategies. Some state building laws have mandated the use of sensor switches in residential bathrooms. As residential users grapple more with issues of energy and convenience, sensors can be expected to appear more often in home applications.

Motion Sensors

Motion sensors, sometimes called *presence detectors*, save energy by switching lighting off when no one is present (i.e., when the sensor does not detect any motion in the space) (see Figure 13.5). They also offer convenience and a sense of security by switching lighting on when the sensor detects motion (avoiding the need for the person to reach around in the dark).

FIGURE 13.5 Motion sensor
Courtesy of Lutron

©Lutron Electronics Co., Inc.

Detection by Passive Infrared Technology

The simplest, most common, and least costly motion sensor consists of a grid of small-scale thermal detection cells that receive infrared radiation from warm objects—like human beings. At any moment, the grid of cells records a pattern, the thermal image of the space within the detection range.

When something warm—like you—moves, the thermal image changes. The sensor interprets the change as presence in the space. When the thermal image remains static for a period of time, the sensor interprets the lack of change as a lack of presence in the space. We discuss how the sensor instructs the lighting to respond under "Sensor Response." follows.

We call this type of sensor "passive infrared" (or PIR) because it passively receives infrared radiation (see Figure 13.6). Note that PIR sensors require a clear line of sight to the space. Blocking a sensor (as with a door or by moving a cabinet in front of it) cuts off the IR radiation so that there is no change in the thermal image, and the sensor "thinks" that the space is empty (whether it actually is or not).

Detection by Ultrasonic Technology

An ultrasonic detector emits an inaudible pressure wave ("sound" you do not hear) and then detects the reflection of the wave in its grid of cells. The sensor interprets a change in the detected pattern as motion and no change as presence.

FIGURE 13.6 PIR sensor
Courtesy of Lutron

The emissions from ultrasonic sensors can travel around partitions, even around corners, so this type of sensor does not need line of sight to the space—an advantage over PIR sensors in some applications.

However, any motion—wind rippling window blinds, for example—appears as human occupancy in the space, which is a disadvantage.

A *dual technology* sensor combines PIR and ultrasonic detection, avoiding the pitfalls of each technology (at a higher cost).

Sensor Response

Motion sensors respond in two basic ways:

1. *Auto on/auto off.* When the sensor detects presence, it turns lights on and keeps them on until it no longer senses presence. After the sensor no longer detects presence, it signals the lights to turn off. Sometimes called an *occupancy sensor*, this control provides both the easy on for security and convenience and the easy off for energy savings.

2. *Manual on/auto off.* Once you turn the lights on yourself, they stay on as long as the sensor detects presence in the space. After the sensor no longer detects presence, it signals the switch to turn off. Sometimes called a *vacancy* sensor, this control maximizes energy savings because lights go on only when you want them on (permitting you to walk into a room without having lights switch on automatically).

Note that both types of sensors detect in the same way; only their response differs.

Sensors can be designed (and programmed at home) to respond quickly when motion is no longer detected or to wait a while before acting. This is called the *time-out period*. Waiting longer (15 to 30 minutes) allows more opportunity for the sensor to detect motion—a friendlier approach than snapping lights off after just a few seconds. (Waiting longer comes at the cost of slightly higher electricity bills.)

Photo Sensors

Photo sensors detect light using a receptor cell of light-sensitive material. Photo sensors are widely used to control outdoor lighting. (A weak reading in the cell signals the switch to turn on; a strong reading signals the switch to turn off.) Indoor usage is growing in commercial and institutional applications for daylight harvesting and balancing.

Daylight harvesting is a strategy for reducing energy consumption with automatic control by reducing electric light when daylight is available. Daylight balancing is a strategy for improving the experience of a transition space by increasing electric light to balance daylight at the perimeter of a space. While these automatic control strategies are not widely used in homes, it is worth understanding their objectives and how they work.

In a daylight harvesting scheme (imagine an office or classroom), the photo sensor reads the combined effect of daylight and electric light in the space. A controller sends a dimming signal to adjust the electric light to maintain a target level of illumination. The electric light dims when daylight is available and increases as daylight diminishes. (Perhaps you noticed that, notwithstanding the name, it is the electric lighting that is being harvested.)

In a daylight balancing scheme (imagine a lobby or boutique exposed to plenty of daylight), the photo sensor reads *exterior* daylight only. When the reading is high (bright daylight), a controller raises the level of electric light so that the interior does not feel so dim in comparison to the bright exterior. As daylight fades, the interior lighting dims so that remains at a comfortable level.

Manual Daylight Control

Of course, you do not need photo sensors to apply daylight harvesting or balancing strategies in a kitchen or large bathroom. You can achieve the same results by creating and *manually* controlling layers of electric light as daylight conditions change.

- *Daylight harvesting.* In a large room, divide the ambient lighting into multiple zones (channels) so that you can regulate the luminaires in a daylighted area separately from those farther away from windows or skylights.
- *Daylight balancing.* Provide sufficient electric lighting on walls and furnishings away that are located away from windows, controlling these luminaires separately. When bright daylight makes these interior surfaces appear dark and uninviting, you can raise the light level (and dim it down as night approaches).
- Using dimmers for variable control over the electric lighting better serves harvesting and balancing strategies.

CONTROL SYSTEMS

So far, our discussion of technology has focused on variations in input (manual controls versus sensors) and effect (how different sources dim). Now let us consider how controls can be linked, or networked, to increase functionality and convenience.

A controls *system* is the term commonly used to describe linked controls. Systems can range from a group of *smart dimmers* that all respond to keypad commands, to LED lamps that can be controlled by a wireless signal from a smart phone or tablet, to a fully integrated home automation system. Making it easy to adjust lighting effects to your desired settings is what a system (rather than stand-alone controls) is all about.

Do you want to link just the controls in the kitchen, or would you prefer to connect controls for kitchen lighting to a keypad at the main entry?

Scope—room or house—is the first question to answer in thinking about systems. Generally speaking, the more you want your controls to do, the more costly the system will be to purchase and install.

It is not surprising that systems limited to one room tend to be simpler and less costly than those connected throughout the home. But whole-house controls can offer significantly greater convenience, including the ability to turn all lights off when leaving, to turn on a "welcome path" through the house when entering, or to set a "return" setting so that you do not enter a darkened home.

Systems for Single Rooms

Imagine an ordinary kitchen. The kitchen opens to a pantry with its own ambient lighting as well as to the dining area and to the main entry hall, three points of entry altogether (see Figure 13.7). The principal activities include food preparation and cleanup, breakfast at the table, coffee or a late snack at the peninsula, and home office work both at the table and peninsula.

The Kitchen
Your lighting design provides task lighting over the countertops, pendants over a breakfast table and peninsula, and ambient lighting throughout (see Figure 13.8). To meet the lighting needs of each activity, you provide a dimmer for each of the four lighting layers. Now, *each* time the kitchen is used, your client needs to find the correct dimmer for each type of lighting and adjust it as desired.

Making Controls More Convenient
Wouldn't it be easier to adjust all of the lighting once for each activity (create scenes), and then simply press a keypad button to recall the appropriate scene? Wouldn't it be nice to neatly label the buttons so that guests can use the kitchen when you are not around? And wouldn't it save time and frustration to place keypads at each of the three entry points so they can be reached regardless of how you are moving through the house? That is exactly what a system does for you!

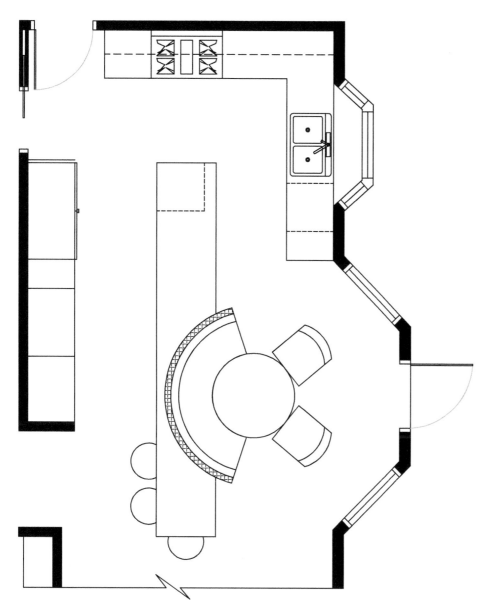

FIGURE 13.7 Kitchen plan

Design by Kim Van Ruskenveld, AKBD, Design Eye Ltd., Edmonton, AB

System Design

This simple system uses four dimmers (one for each layer) and three keypads, one for each entry. Each keypad has five buttons (see Figure 13.9) that can be customized by name, such as "Coffee," "Breakfast," "Dinner," "Homework," and "Off"—one for each scene.

Each button tells all four dimmers to recall the specific setting that is part of the scene you want to create. A network signal wire or wireless transmitters and receivers might be used to connect the components to each other.

The dimmers can be ganged together in one location or distributed so they are near the lighting they control. Ganging often looks cleaner. Distributing makes it easier to temporarily override any scene to make a lighting layer brighter or dimmer.

Wireless communication is much more flexible than wired connections and allows for less costly remodeling. (There is no need to run new signal wires.) But wireless carries a higher price and, depending on product quality, may be less reliable.

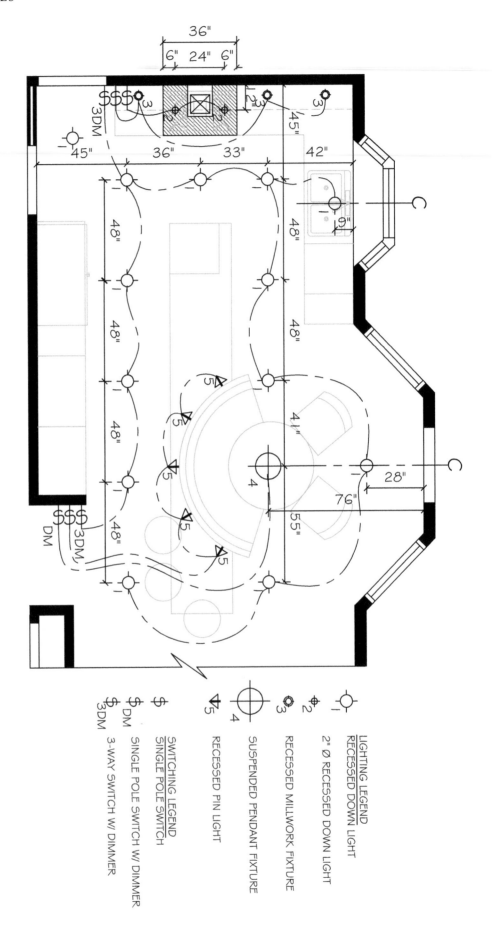

FIGURE 13.8 Kitchen plan—lighting and basic control

Design by Kim Van Ruskenveld, AKBD, Design Eye Ltd., Edmonton, AB

To determine what kind of system suits the project, you have to have an idea of what is required. Following are some points that must be considered.

- The number of *dimmers* depends on the number of lighting layers and control channels you want.
- The number of *keypads* depends on how many control locations you want.
- The number of *buttons* on the keypads depends on how many different scenes or settings you want.

The Bath

Although many bathrooms are simple enough to light and control with two or three stand-alone devices, other spaces are larger and benefit from a control system (see Figure 13.10). These spaces host a wide variety of activities, ranging from routine grooming, to luxuriating in a spa, to exercising. Multiple layers of lighting, each independently controlled, support the diverse activities and client preferences.

The same type of system and design approach sketched for our kitchen works in this larger and more complex bath area.

Controlling Adjacent Spaces

The preceding examples focused on self-contained spaces: single rooms. Room control systems can extend to control adjacent areas as well. A kitchen might open onto the dining area without intervening walls; the pantry might be an alcove off the main kitchen space; the bath might be en suite to the master bedroom or dressing area.

In these cases, you need to consider systems offering greater flexibility in the number of dimmers, keypads, and buttons. Conceptually, however, designing controls and settings is the same process.

Systems for the Entire House

You typically control a single-room system from within the space or from a visually connected adjacent space. To control more than one room and to control spaces from remote locations (such as the front or rear doors, a different floor, or the master bedroom), you want to use a system designed for an entire house.

For house control, you begin with a control (switch or dimmer) for each lighting layer. As with room control, these controls are then connected to keypads that control one or more switch or dimmer. As with single-room systems, the connections can be wired or wireless.

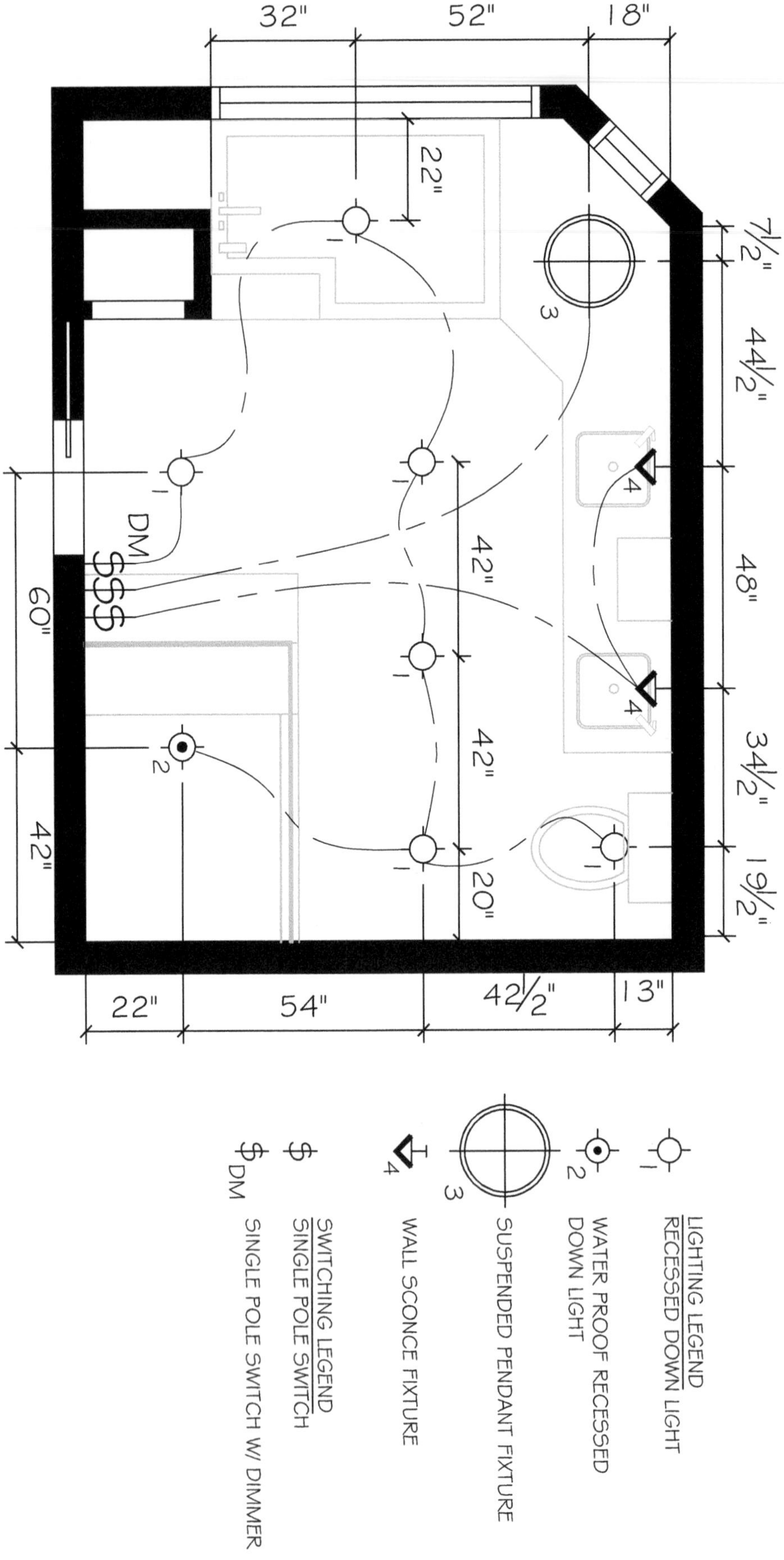

FIGURE 13.10 Bathroom lighting plan and system

Design by Nicole Campbell, Design Eye Ltd., Edmonton, AB

Some manufacturers offer products that can be used to control either a single room or an entire house. These products can simplify design, installation, and troubleshooting.

Strategies

Systems for controlling an entire house are very powerful. You need to decide what you want the system to do; otherwise, it can become burdened with more cost and complexity than is really needed.

Some strategies are listed next.

- *Entry and exit.* Arrange controls to turn on selected lights in specific rooms to provide a comfortable and secure entry. Turn off all lights upon leaving. Or turn off most lighting and leave security lights on.
- *Paths of light.* Arrange controls to turn on a path of light from one area of the house to another (e.g., entry to kitchen or bedroom or master bedroom to kitchen).
- *Entertainment.* Arrange controls for scenes in multiple areas. This strategy works for houses with both open-plan and traditional layouts.
- *Lights out.* Turn lights out throughout the house (or leave a few on) from a keypad in the master bedroom. This is especially useful if the kitchen is on a different floor or if children tend to leave lights on in their bathrooms.
- *Emergency.* Turn all lights on (or just strategically located ones) to provide light and an alert in case of emergency.
- *External connections.* Turn lights on and off from a remote signal, such as your phone, computer, garage opener, audio-visual controller, or security system.

WIRELESS CONTROL OF LED LAMPS

The newest form of controllable lighting features wireless control of one or a group of special LED lamps. Using an ordinary wireless Internet router and a proprietary hub, you can control these special light sources from your smart phone, tablet, or a dedicated remote.

Any portable or fixed luminaire with a medium-base socket can be fitted with these lamps, which typically pass the wireless signal from receiver to receiver, allowing the control to span an entire house.

Some products simply dim white light; others can change color as well. A range of form factors, such as tubular or tape, is becoming available.

Notwithstanding the many limitations of this emerging technology (notably light output, beam control, and color choice), it offers a practical and cost-effective approach to adding control to existing lighting systems.

APPLYING CONTROLS

Control over your lighting is your responsibility as the designer; do not delegate it to a contractor or even an engineer.

1. Start thinking about controls at the beginning of the design. Develop the controls plan based on the activities and users in the space and the lighting effects you are creating.
2. Assign a control for each layer of lighting. Divide layers into multiple channels if the lighting serves multiple areas in the space. Control different light sources on separate channels to ensure compatibility with the control equipment.
3. Decide whether you want manual or automatic input to the control and on/off or variable (dimmed) effects from the control.
4. Select controls, particularly dimmer, that are compatible with the type and magnitude of the load.
5. Decide if stand-alone controls will suffice or whether a system better meets your objectives.

6. For a system, identify the scenes, paths, or other strategies you want to create and select scene-control equipment accordingly.

7. Locate controls near entries and other convenient places. Use remotes to handle multiple locations.

Lighting Controls

This exercise involves an inventory of lighting controls in a space and evaluation of their effectiveness. You can use your own home or any other space (ideally a kitchen or bath). Use your journal to record your observations.

Create a list of all of the lighting controls in the room or space you have been working (or a new one, if you prefer). Describe each one by the following:

a. User interface (manual or automatic)
b. Effect (on/off or variable)
c. Linkage (stand-alone, three-way, or networked)
d. Load: type of source and total wattage controlled
e. Location
f. Is each layer of light in the space on a separate control?
g. If not, what limitations does this impose?
h. How easy is it to use each control? Consider its location, method of adjustment, and feedback to you.

SUMMARY

All lighting needs to be controlled. The question is how: how you *interact* with the control, how the control *adjusts* the light, and how controls are *linked*. Switches turn lights on and off; dimmers vary light output (and, for incandescent sources, change the color of the light). Controls must be compatible with the loads—both type and magnitude—they handle. Systems offer more convenience and sense of security than stand-alone controls. For the best results, controls should be considered at the beginning of lighting design, and their complexity and cost should be aligned with design and client objectives.

REVIEW QUESTIONS

1. What is a load? What is the difference between a dimmer and a user interface? What is a preset, and why is it valuable (See "Language of Controls" pages 195–197)

2. What are the benefits and limitations of dimming incandescent sources? (See "Dimming Incandescent Loads" pages 198–199)

3. What are the benefits and limitations of dimming LED sources? (See "Dimming LED Loads" page 201)

4. What are the benefits and limitations of dimming fluorescent sources? (See "Dimming Fluorescent Loads" pages 201–202)

5. What are the benefits of a system compared to stand-alone controls? (See "Control Systems" pages 206–211)

Design Development

Development turns a lighting concept into a practical, buildable lighting design. The development phase pins down the location and specification of luminaires and controls. It resolves questions of architectural integration, code compliance, and lighting requirements. With the completion of design development, you are ready to document the design so that it can be properly bid and constructed.

Learning Objective 1: Locate luminaires in kitchen and bath designs so as to implement lighting design concepts.

Learning Objective 2: Select specific luminaires of different types.

Learning Objective 3: Determine the lumen rating of lamps and luminaires to meet lighting requirements.

Learning Objective 4: Select and locate controls.

DEVELOPING THE DESIGN

You began the lighting design process with a program reflecting the needs and objectives of the client: functional, aesthetic, and economic. This evolved into a list of lighting criteria for the design. Considering the lighting criteria in the context of the architectural and interior design and the fundamentals of light and vision, you created one or more lighting concepts.

With the schematic design approved, it is time to develop the concept in detail—that is, to locate, select, and control specific luminaires.

In design development, you will:

- Use detailed lamp, luminaire, and controls specification information from the manufacturers with which you are most comfortable.
- Locate the luminaires *specifically* in the space in order to distribute the light to the targeted surfaces. Location involves distance from work surfaces and vertical surfaces as well as arrangement of all luminaires together.
- Specify the luminaires, moving from a concept (direct lighting over a peninsula) to a specific luminaire type. This includes determining the light source, distribution and control of the light, mounting method, and appearance as well as ensuring that the choices comply with relevant codes and regulations.

- Size each light source. Here you need to determine quantity of light in the lamp and luminaire that will fulfill the design criteria. For example, do you need 600 lumens or 1000 lumens?
- Control the luminaires. You will implement the control plan by identifying the luminaires to be controlled and selecting specific switches, dimmers, sensors, and other interface devices needed to accomplish the plan.
- Verify that your design complies with the applicable codes and regulations. This is a check. Presumably, you are already aware of the code requirements and have not located or specified equipment so as to violate any codes or regulations.

Note that the process does not simply proceed in a straight line. Instead, the decisions overlap to a degree, so the development process tends be iterative and holistic, just like other aspects of design.

CODE COMPLIANCE

Although we just listed verifying compliance at the end of your design development process, we are going to begin with codes because they affect so much of the specifics of lighting equipment.

Two types of codes and regulations chiefly govern residential lighting: electric and energy. Although standards are created nationally, the legal codes are written, enacted, and enforced within state or, in some cases, municipal jurisdictions (provinces in Canada). Local rules may be more stringent than national standards but not more lenient. And the local inspector's interpretation typically prevails.

Electric Code

The national standard for electrical installation in the United States is known as the National Electric Code (NEC). Technically, the NEC is a misnomer; it is not a code until enacted *locally*. The NEC is published by the National Fire Prevention Association. The Canadian Electric Code is published by the Canadian Standards Authority (CSA).

Broadly speaking, the installing electrical *contractor* is responsible for most electrical code compliance. Nevertheless, you should not specify lighting that does not comply. Some common areas to understand include these:

- Luminaire labels
- Type IC downlights
- Dry, damp, and wet locations
- Low-voltage wiring

Luminaire Labels
To ensure that lighting equipment can be installed in accordance with the NEC, Underwriters Laboratories® (UL) establishes *product standards*. Similar standards for Canada are written by CSA.

UL and other independent laboratories review lighting products for compliance with UL and CSA standards, listing those products that have been reviewed according to specific procedures. Some standards require specific construction; others require testing, typically to avoid overheating electrical wiring or combustible materials.

UL (and other labs) furnish labels to manufacturers, who place them in luminaires so that installers and inspectors can verify that the product has been listed.

Labeling requirements (note: all data are not on a single label) also contain important information—notably the *maximum wattage and type of light source* that can be installed in the luminaire (see Figure 14.1). The label also indicates where the luminaire is *electrically suitable*, such as for dry, damp, or wet locations. (Note that such a label does *not* imply that the luminaire is functionally suitable.) As a designer, typically you can find the label information in the luminaire specifications.

FIGURE 14.1 Luminaire label
Courtesy of American Lighting Association

Type IC Downlights

Recessed luminaires embedded in thermal insulation tend to heat up more than those installed away from insulation, possibly affecting electrical insulation and connections.

Energy codes typically require thermal insulation at the exterior layer of the home. Luminaires intended for such application must pass specific tests. Those that do are labeled type IC (for insulated ceiling). Thermal considerations can limit the wattage (and amount of light) in the luminaire and the depth of the light source.

While ceiling construction often affects decisions about whether to recess luminaires or not, the presence of insulation largely governs the type of recessed luminaire used. These luminaires also must be airtight as well.

Dry, Damp, and Wet Locations

Since moisture can affect electrical safety, luminaires are labeled as to the appropriate location: dry, damp, or wet. Most inspectors consider showers and spas as wet locations; the rest of the bath area would generally be considered a damp location. Note that this listing only considers electrical suitability; it does *not* imply functional suitability.

Low-Voltage Wiring

Low-voltage lighting systems—those with transformers—have two basic classifications:

1. *Class I transformers handle more than 60 watts (W).* These larger devices require a fuse, use larger conductors, and typically require a concealed, accessible location, such as closet or cabinet.
2. *Class II transformers handle less than 60 W.* These small devices do not require a fuse, use smaller conductors, and typically can be exposed (like a typical computer power supply). Plug-in types are generally Class II, affording installation flexibility.

Energy Codes

With the growth in national concern over energy consumption, there has been an increase in energy codes governing residential construction, including lighting.

Although the *minimum* national requirement for energy codes does not cover residences, some states have adopted more stringent codes. This includes California's Title 24. Most others follow the International Energy Conservation Code (IECC), which has some residential lighting provisions.

These codes impose two typical requirements:

1. Minimum amount of high-efficacy lighting equipment
2. Use of qualified controls

High-Efficacy Luminaires

As used in California's Title 24, *high-efficacy luminaires* include those that are dedicated to fluorescent or light-emitting diode (LED) sources. Luminaires with screw-shell sockets (those that can accept incandescent sources) do *not* qualify.

Title 24 (2008) requires:

- *At least 50 percent of the power (watts) used for kitchen lighting be in high-efficacy luminaires.* Recognizing that high-efficacy luminaires produce two to five times the lumens per watt of incandescent luminaires, this provision effectively mandates the use of energy-efficient equipment as the predominant (but not total) light source in the kitchen. In residential kitchens, the installed lighting power of electrical boxes finished with a blank cover or where no equipment has been installedshall be calculated as 180 watts of low-efficacy light per box. Inside cabinet lighting is capped at 20 watts per linear foot.
- *At least one luminaire in the bathroom must be high efficacy.* All other luminaires in bathrooms must either be high efficacy *or* controlled by a vacancy sensor. Laundry, utility, and closet areas that may be part of a kitchen design must be high efficacy or controlled by a vacancy sensor.

To verify compliance in a kitchen design, you calculate the wattage of all luminaires in the kitchen and determine what percentage of the wattage is represented by fluorescent or LED sources (excluding screw-base lamps).

High-Efficacy Light Sources

The IECC (2012) takes a different approach. It requires high-efficacy light sources in at least 75 percent of permanently installed luminaires *throughout the home.* (Earlier versions were 50 percent and less.)

High-efficacy sources are defined as:

Source Wattage	Minimum Lumens per Watt
≤ 15 W	40
15–40 W	50
≥ 40 W	60

This definition permits the use of screw-base compact fluorescent (CFL) and light-emitting diode (LED) lamps in incandescent luminaires (which Title 24 specifically prohibits). Judging by the performance of current LED lamps and luminaires, most will meet the IECC requirement.

To verify compliance, you compare the number of luminaires with high-efficacy sources to the total number of luminaires used in the home. (Note this is *not* the number of *different* luminaires.) Since you may be responsible for only part of the home design, ensuring code compliance (where IECC applies) will require specific communication with the rest of the design and construction team.

LOCATING LUMINAIRES

Creating a beautiful and functional residential design depends more on locating the lighting than on calculating how many footcandles are delivered on average.

If you locate luminaires where you want the light, you will have little difficulty in sizing the illumination. If the luminaires are located inappropriately, however, adjusting the amount of light rarely solves the problem.

Using location as your first principle may lead you to use more luminaires than you initially considered. This is not a problem. You can always adjust your design to eliminate those luminaires that, upon review, seem to be duplicating the light distribution.

Moreover, more luminaires with less light in each one generally produces more comfortable and effective lighting than fewer luminaires with more light in each one.

Downlights

If you want to provide *direct* lighting over a surface, locate luminaires over that surface. But how many downlights do you need, and how should they be spaced?

Spacing and Beam Spread

The arrangement of luminaires in the ceiling—specifically how the arrangement appears—is an important consideration. Downlights can be comfortably spaced farther apart (4–6 feet on center) in larger rooms and higher ceilings. Closer spacing (3–5 feet on center) works better for smaller rooms and lower ceilings.

Wider beams are needed for downlights spaced farther apart and *closer* to the target surface.

Narrower beams will work for downlights spaced closer to each other and farther from the target surface (a higher ceiling).

You can easily verify that your choice will deliver uniform illumination across the surface with the luminaire's spacing ratio or spacing criteria, which are derived from the photometry of the luminaire.

Applying the Spacing Ratio

The spacing ratio (S.R.) tells you the *maximum* spacing that will give you uniform lighting on a target surface with a *specific* luminaire.

1. Determine the mounting height from the luminaire down to the target surface. For example, a downlight installed in a 9-foot ceiling and lighting countertop at 36 inches has a mounting height (M.H.) of 6.0 feet. Over a 28-inch table, M.H. equals 80 inches or 6.7 feet. And over a 42-inch counter, the M.H. equals 5.5 feet. (Expressing M.H. in decimals makes the math easier.)
2. Find the S.R. for the luminaire you plan to use. The number, which typically ranges from 0.4 to 1.6, differs according to the optics of the luminaire and is found with the luminaire's technical data.
3. Multiply the S.R × M.H. to find the maximum spacing between luminaires.

Example: Suppose we are using a downlight with a narrow beam (S.R. of 0.4) over the 42-inch counter. The maximum spacing between luminaires for even illumination is 0.4 × 5.5 or 2.2 feet or 26 inches apart.

Now suppose that when you look at this spacing on your kitchen plan, you notice that it feels too close. You can change your luminaire selection to one with wider light distribution (or accept the uneven distribution of light, which may not be objectionable in a social area). Which option would you choose?

Downlights near Cabinets and Walls

The beam of light from a downlight creates a scallop-shape patch of brightness on nearby vertical surfaces, such as cabinet fronts and walls. (see Figure 14.2). The closer the wall or

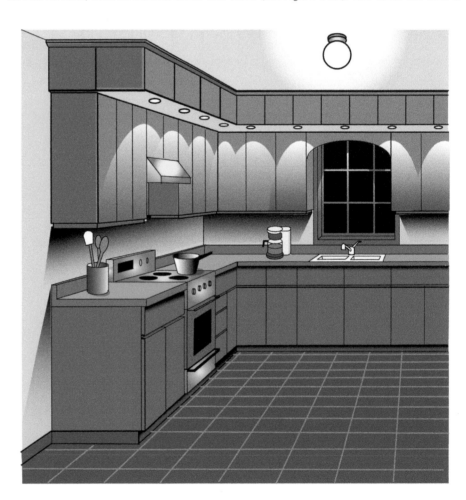

FIGURE 14.2 Scallop effect on cabinets created by recessed downlights in the soffit

cabinet is to the luminaires, the brighter the pattern and the higher on the surface it falls.

Called scallops, this lighting effect draws attention, whether intended or not. Thus, you should arrange downlights so that scallop patterns fall where you want them to appear.

To arrange downlights near cabinets, use an elevation of the cabinetry (which makes it easier to visualize the relationships) as well as your floor and reflected ceiling plans. Locate the luminaires so they relate visually to the cabinet elements and in particular the centerline of the doors.

Downlights work most effectively when they are 12 to 24 inches from the face of the cabinets. Here, they can illuminate both the cabinet interior and the front of the counter. Closer, and the brightness of scallops will overwhelm; farther away, and a person at the counter will create a body shadow on the work surface.

Where downlights will be seen against a wall or door, try to locate the luminaires symmetrically. This way, they will not draw as much attention to themselves. Note that symmetry is less important in smaller spaces with abbreviated sight lines.

Downlights over a Sink

A single downlight centered over a bathroom sink looks well organized, but it is completely wrong! Located directly over the basin, a direct light creates deep, harsh facial shadows that make it harder to apply makeup, to shave, and to perform other grooming tasks, unless combined with sconces on both sides of the mirror, so light is approaching the face from three directions.

Downlights should be placed on either side of the basin so they can light across rather than directly down. Locate the luminaire as close to the mirror as possible (away from the face) so that the light softens features. With multiple basins, locate the downlights to the side and between.

A downlight directly above a kitchen sink creates hand shadows when working. Reflected light from the sides of adjacent cabinets mitigates the problem, however.

In both kitchens and baths, you can reduce shadowing by using a luminaire with a wide, soft beam rather than one with a narrow, sharp beam.

Lighting from under Cabinets

Lighting for kitchen counters typically—and effectively— is located under the cabinets immediately over the countertops. But where should it be placed: near the front of the cabinet or the back? The best approach may depend on the materials used for the countertop and backsplash and the primary tasks in the kitchen (see Figure 14.3).

Lighting under the Front of the Cabinet

For most materials and most activities, under-cabinet task luminaires should be located at the front of the cabinet, where the light will fall most directly on the primary task surface (see Figure 14.4). This is the typical solution.

Lighting under the Back of the Cabinet

If the kitchen serves a more social, rather than working, role, and a distinctive backsplash will benefit from a grazing light, you may prefer to place the lighting at the back of the cabinet.

Fascias and Shielding

Although under-cabinet luminaires may seem to be out of sight when you are standing at the counter, they can be uncomfortably glary when viewed from farther into the room, especially when seated.

Apply a fascia to shield the lighting, and ensure that it is deep enough to prevent view of the luminaire when seated. The size of the fascia depends on the size of the luminaire, its position, and the viewing angles. LED tape light or other shallow linear luminaires require a smaller fascia than fluorescent or incandescent task lights.

FIGURE 14.3 The under cabinet lighting in this kitchen provides task lighting for the counters and draws attention to the brick detail of the backsplash.

Design by Carl Bruen, CGR; Co-designer Robin Bruen and Debbie Kerr, CKD, Bruen Design Build, Inc., Morristown, NJ
Photo by Wing Wang

FIGURE 14.4 Under-cabinet lighting should be located at the front of the cabinet.
Courtesy of Hafele

FIGURE 14.5 Interior cabinet lighting should be located near the front of the cabinet.
Courtesy of Hafele

Be sure to consider the view into the side of cabinets, which may expose the entire length of the luminaire.

Lighting inside Cabinets

Many installers and designers locate lighting inside cabinetry in the middle of the cabinet, but that is not the best approach. If you want to illuminate the objects inside a cabinet, place the luminaires as close to the front of the cabinet as possible, as seen in Figure 14.5. From this location, most of the light will fall on most of the objects. (Lighted from the middle of the cabinet, the objects toward the front—the most important ones—will be in shadow.) Another approach is placing linear lighting at the hinge side of the door facing inward (see Figure 14.6). This allows for good illumination of the materials inside.

Coves

Indirect lighting from a cove can deliver comfortable, well-shielded illumination along with architectural integration. Cove lighting can serve as the primary ambient illumination in a kitchen or bath, or it can provide a gentle architectural accent to highlight ceiling, wall, and cabinet details.

Ambient Illumination

Fluorescent strip lights perform well as ambient sources, and they are very economical. LED luminaires are an alternative to the fluorescent.

Generally speaking, the smaller the cove, the better it looks; the larger the cove, the better the illumination. Where you want to provide generous ambient illumination, size the cove appropriately.

Ideally, a fluorescent cove measures at least 12 inches deep (from front to back) and 18 to 24 inches to the ceiling. This allows light to escape the cove and spread across the ceiling. When a cove is built on top of cabinetry, these dimensions are practical; otherwise, you may find them too large for the space.

FIGURE 14.6 Interior cabinet lighting can also be installed along the side of the cabinet.
Courtesy of Hafele

The fascia enclosing the cove need be tall enough only to conceal the lighting equipment. The top of the light source should be precisely even with (and no higher than) the top of the fascia. The interior of the cover should be painted white for good reflectance.

In a smaller space, such as a bathroom, indirect coves can be compressed to half the dimensions just indicated because a smooth spread of light across the ceiling is not important.

Not sure you have sufficient space for an effective cove?

- Fluorescent strip luminaires with T5 lamps, which can be half the height and width of T8 versions, are a good choice where the cove dimensions are restricted.
- LED coves can be smaller as well because a directional luminaire can be arranged to distribute light into the room.

Accent Illumination

Coves and pockets intended to highlight an architectural feature, such as molding or a toe-kick, can be sized to the light source. Space is not as important as it is with ambient lighting.

Fluorescent sources are typically too bright for accent applications. Consider LED luminaires for smooth, white light. Dimmed xenon or halogen strip lights make sense where you want to create a warm and cozy atmosphere. Xenon linear systems are starting to be replaced by LED. LED is a faster installation and does not use as much energy, nor does it generate as much heat.

These small-scale luminaires typically use a remote transformer or power supply, which may not fit into a tightly dimensioned cove. Be sure to provide a location for these remote components so you can access them for maintenance or replacement.

Socket Shadows

In an ambient cove, the dark ends of fluorescent lamps can create "socket shadows" against the back wall of the cove. Overlapping the lamps (sometimes called *staggering*) by 4 to 6 inches alleviates this problem.

In an accent cove, the spaces between individual light sources (incandescent or fluorescent) create similar shadows. Where possible, use a luminaire with sockets spaced close together, and position it as far as practical from visible surfaces. The new LED strip products are continuous and eliminate the shadowing between the source lamps (see Figure 14.7).

FIGURE 14.7 This bathroom includes several types of lighting applications. Note the ceiling, mirrors, and alcoves.

Courtesy of Kichler

Pendants

Locating pendants on the lighting plan is fairly simple; they relate to the surface over which you mount them. The critical question is determining the suspension height so that people can see each other across the pendant while not experiencing glare from the light source. (This is not a problem, of course, for uplight pendants.)

- *Over a breakfast table.* A comfortable guideline is suspending a pendant 26 to 30 inches above the table. Ceiling height and table size may influence this dimension a bit.
- *Over a peninsula counter.* Since people may be seated or standing, keep the bottom of the pendant above *standing* eye level—and ideally 6 inches higher still.

Closet Lighting

Can't find what you want in the pantry? Having a hard time picking out that medicine bottle? Think about better lighting for closets.

Shallow closets often can be effectively lighted from luminaires in the main kitchen or bath area. Otherwise, consider a simple luminaire mounted inside the closet and over the door. Deeper closets may require locating the luminaire in the middle of the space. Arranging the luminaires so they cover the storage areas is more important than providing a lot of light from each one.

With appropriate color quality, LED and fluorescent lighting (and door-activated switches) are good approaches for lighting within closets.

The NEC is concerned about hot electrical devices (luminaires) in close proximity to the combustible materials in closets. Hence the following provisions:

- All incandescent luminaires must be enclosed (no bare lamps).
- Surface-mounted incandescent luminaires must be at least 12 inches from the edge of any shelf or hanging area.
- Recessed incandescent luminaires must be at least 6 inches away from the edge of any shelf or hanging area.
- Fluorescent and dedicated LED and CFL luminaires must be at least 6 inches away from the edge of any shelf or hanging.

Some jurisdictions may limit lighting in closets further. For example, California's Title 24 requires high-efficacy lighting (see below) in closets over 70 square feet.

Bath Lighting

Bath areas typically divide into several zones, sometimes separately enclosed, sometimes in an open plan. Areas with separate enclosures (e.g., toilet, walk-in closet) need separate luminaires.

Luminaires with glass or plastic diffusers—recessed or surface mounted—spread light pleasantly. Ordinary downlights, which cast light downward, leave upper walls dark, make the space feel gloomy, and make it harder to see into upper shelves.

At the Vanity
Lighting that surrounds your face flatters with soft illumination and few shadows. Lighting from a single point, however, creates both shadows and glare.

- *Lighting from all sides of the mirror.* This arrangement not only favors facial rendering; it also reduces glare by distributing brightness over a larger area (see Figure 14.8).
- *Lighting from both sides of the mirror.* Linear luminaires placed vertically on either side provides both flattering and comfortable illumination and have the benefit of simplicity (see Figure 14.9).
- *Lighting above the mirror.* This is a typical approach. While it is economical, it is less effective than the previous ones. Use the longest luminaire that will fit, and choose one with a diffusing enclosure to soften the light (see Figure 14.10).

FIGURE 14.8 Lighting from all sides is built into these electric mirrors.
Design by designstudio ltd., courtesy of Electric Mirror
Photo by Eric Laignel

FIGURE 14.9 Lighting from both sides of the mirror
Design by Leslie Lamarre, CKD, CID, CGBP, codesigner Erika Shjeflo, CID, TRG Architects, Burlingame, CA
Photo by Bernard Andre Photography

FIGURE 14.10 Lighting from above the mirror

Design by Corey Shannon Klassen, CKD, CBD; codesigners Ian MacDonald and Scott Lumby, Corey Klassen Interior Design, Vancouver, BC
Photo by Jason Karman

For long mirrors over a multi-sink vanity, you can mount luminaires to the mirror between each basin or run several together across the top.

Reflective surrounding surfaces also help to minimize shadows and glare by distributing light effectively around the face and from below.

In the Shower
If translucent glass or a shower curtain encloses your shower and there is ample ambient illumination in the rest of the bathroom, you may not need a separate shower light (see Figure 14.11). Most detailed visual tasks take place at the vanity.

Where ambient light does not adequately penetrate the shower:

- *Utilize light from above* with a recessed downlight designated for use in the shower. Typically, this will feature an enclosed and gasketed diffuser and carry a wet location listing.
- *Utilize light from the side* with luminaires mounted in or on the walls. You will need to ensure the fixture is rated for this type of wet application.

Luxuriating in the Tub
An elegant bathtub or spa deserves lighting that enhances enjoyment, relaxation, and a moment of luxury (see Figure 14.12). Two important lighting design tips for the bath are:

- Locate overhead lighting so that does not shine directly into the eyes of someone below, and use well-shielded luminaires.

FIGURE 14.11 Shower without its own light source

Design by Anastasia Rentzos, CKD, CBD, Andros Kitchen & Bath Designs, Mississauga, ON
Photo by Averill Lehan

- Lighting around a spa or whirlpool bath (tubs with electrical motors) needs an additional safeguard. Within 8 feet of the water, the luminaire must be wet location listed and not have any exposed metal parts that might conduct electricity.

Accent Lighting

Paintings, photographs, and other graphic works of art mounted on a wall can be illuminated using three basic techniques:

1. *Light the entire wall and the art together.* Contrast between the paint or photograph and the color of the wall enables the work to draw attention.

FIGURE 14.12 Lighting can enhance the bathing experience.
Design by Tammy MacKay, AKBD, and Christine Pandur, Design Eye Ltd., Edmonton, AB

2. *Light the entire work, with some illumination spilling over the frame.* The work should center in the scallop patch of light on the wall; otherwise, the asymmetry will be the most prominent visual element in the space.

3. *Confine the light to the interior of the work* so that it appears to glow by itself.

Objects appear most natural when lighted from an angle of about 30 degrees from vertical. Accent luminaires should be located so they can be aimed at this angle (see Figure 14.13).

To locate a recessed or track-mounted adjustable luminaire for lighting art on a wall, follow these four steps:

1. Identify the center of the work to be lighted. This is the aiming point.
2. Calculate the distance from the aiming point the ceiling.
3. Multiply that distance by 0.55, which gives you the distance from the wall.
4. If the desired location cannot be used (e.g., due to an obstruction), move a recessed luminaire slightly closer to the wall rather than farther away.

Objects in niches or cabinets need light from in front (not centered in the niche or cabinet); otherwise, they will be in shadow. See "Lighting inside Cabinets."

SELECTING LUMINAIRES

Having located luminaires to realize your lighting concept, you are ready to nail down exactly which ones you want to use.

FIGURE 14.13 Locating accent lighting

Design by Tammy MacKay, AKBD, and Christine Pandur, Design Eye Ltd., Edmonton, AB

We will look at six basic luminaire types (coves were covered earlier):

1. Recessed downlights
2. Under-cabinet task lights
3. Decorative luminaires
4. Fluorescent luminaires
5. LED luminaires
6. Adjustable accent luminaires

Recessed Downlights

You can select recessed downlights in five basic steps:

1. *Quality.* Product quality (functionality and detail) goes along with equipment cost.
2. *Light source and effect.* Consider color, beam, dimming, and sustainable operation (energy and life).
3. *Size.* This is a matter of the *quantity* of light desired and the physical dimensions. ("Small is beautiful.")
4. *Ceiling conditions.* Is there insulation? How deep is the plenum?
5. *Aesthetic options.* Decorative materials, finishes, trim details to compliment the entire project.

TABLE 14.1 Light Sources for Downlights

	Incandescent	**LED**	**CFL**
Color	Warm, familiar tone	Choice: warm to cool, but consistency and color rendering index (CRI) can vary	Choice: warm to cool, but CRI may disappoint
Beam	A lamps provide wide beams; reflector lamps provide medium to wide beams	LED lamps offer the same options as halogen; LED luminaires provide medium to wide beam	Medium to wide beam only
Dimming	Simple: all sources dim, color warms	Complex: issues of compatibility; most do not warm color	Limited: requires special lamp or ballast, which increases cost
Sustainability	Poor: low lumens per watt and short life	Best: high lumens per watt and long life	Medium: both lumens per watt and life

Quality

Downlights fall into three basic categories: specification grade, residential grade, and builder grade. The names are descriptive and do not necessarily correspond to manufacturers' designations, which vary from brand to brand.

Specification-grade luminaires are basically products specified for commercial and institutional applications, adapted for residential construction. They offer the best performance, cleanest appearance, and most choice. But these products can carry a price that two to three times that of lower-quality versions. Where high-end products are used throughout the kitchen and bath and the budget is ample, this grade should be the first choice.

Residential-grade luminaires are developed for both residential and light commercial use and are chosen by most design professionals and higher-quality distributors and contractors. Versions are available for all of the applications in a typical home. As you would expect, some brands do a better job in one attribute or another. Without budget guidance, this is a good place to start.

Builder-grade luminaires meet the minimum requirements for residential application. They are often the choice for lower-priced homes, especially where there is no direct homeowner involvement. Given the compromises in the product, this might be considered a last resort unless otherwise directed.

Light Source and Effect

Table 14.1 compares light sources for downlights across several attributes. Depending on the parameters of the design, one will best implement the design intent.

Size

The appropriate size for downlights depends on the quantity of light needed and the scale of the space in which the luminaire will be used. Table 14.2 looks at three typical aperture sizes for downlights and the range of light output they can handle. As you can see by the range of lamp lumens, some applications (lighting requirements) could be served by all sizes of downlights; others, by only one.

TABLE 14.2 Typical Aperture Sizes

Nominal Aperture	**Lamping**	**Lamp Lumens**
3 inches	R20, PAR20, MR16	400–800
4–5 inches	A19, BR30, PAR30, MR16, CFL	600–1200
6–7 inches	A19/21, BR30/40, PAR30/38, CFL	600–1800

Ceiling Conditions

As noted earlier, the presence of ceiling insulation determines whether you need to use a type IC luminaire. The thermal limitations imposed by insulation limit lighting choices, and this should be addressed as early as possible to avoid last-minute modifications.

Shallow conditions also may limit choices, and this too should be confronted early—perhaps recessed luminaires are not the appropriate choice after all. Most residential-grade product lines offer housing and trim conditions that fit in a shallow opening.

New construction allows installation of the downlight housing before the ceiling is enclosed. Where the ceiling already exists, so-called remodeler downlights may enter through a ceiling cut-out the size of the diameter (easy to do) or through a larger patch (more difficult and costly).

Aesthetic Options

Theoretically, the objective of using recessed luminaires is a quiet ceiling, where the equipment simply "disappears" or at least is inconspicuous. But opinions differ as to how best to accomplish the goal.

A clear aluminum finish on the interior performs optimally—both when the luminaire is illuminated and when it is not. The luminaire may appear to be on even when it is not. This can be a desirable look but also can create glare and be distracting. Some designers prefer a white finish, which blends easily into the ceiling when the luminaire is off but can be uncomfortably bright when it is on. A black finish looks very quiet when the light is on but may feel like a dark hole when it is off.

If you want a more ornamental appearance, you can choose from an array of special finishes and glass or metal attachments.

Under-Cabinet Task Lighting

There are two key considerations in selecting luminaires for under-cabinet task lighting. First is the light source (incandescent, LED, or fluorescent). Most under-cabinet task luminaires are designed around one of these specific light sources.

The second consideration when selecting luminaries for under-cabinet task lighting is luminaire form, which may be either linear or discrete ("puck" style). Linear task luminaires can illuminate the entire length of the work surface (or backsplash) uniformly. With an effective fascia, they are largely concealed.

Discrete luminaires, in contrast, do a better job of providing highlights and may be more attractive where it is not possible to use a concealing fascia.

Table 14.3 considers the source and luminaire together.

TABLE 14.3 Comparing Light Sources for Undercabinet Illumination

	Halogen	**LED**	**Fluorescent**
Color	Warm, familiar tone	Choice: warm to cool, but consistency and CRI can vary	Choice: warm to cool, but CRI may disappoint
Beam	Some lamp/fixture combinations offer directional control or reflectors that aim the light; others do not	LED "tape" emits light in about 180 degrees; some luminaires provide even better control	Lamp emits light in all directions; luminaire offers only limited control with lots of spill
Dimming	Simple: all sources dim, color warms	Complex: issues of compatibility; most do not warm color	Limited to larger luminaires; requires special ballast, which increases cost
Sustainability	Poor: low lumens per watt; short to medium life	Best: high lumens per watt and long life	Medium: both lumens per watt and life
Luminaire Form Factors	Linear and discrete; flexible layouts, small profile; often hot to the touch	Linear and discrete; flexible layouts; small profile; tape can be ultra-compact	Linear only; fixed lengths; relatively bulky profile

FIGURE 14.14 It is important to coordinate the finishes on the lighting fixtures in relation to the other fixtures in the space.
Design by John Sylvestre, CKD, Sylvestre Construction, Inc., Minneapolis, MN
Photo by Karen Melvin Photography

Decorative Luminaires

While style is the chief criterion for selecting decorative luminaires, here are a few other considerations.

Kitchens and bathrooms tend to have an assortment of style treatments, materials, and finishes: tile, glassware, hardware, appliances, even fabrics. Most decorative luminaires use different materials and finishes than are used to furnish these spaces. A polished nickel finish on a faucet set may not look the same as the polished nickel on the back plate of a wall sconce. This can make style coordination a little tricky. Sometimes simpler luminaire designs are easier to integrate than more ornate ones (see Figure 14.14).

Light Sources

In most applications, general service LED and CFL lamps can effectively replace medium-base incandescent general service sources. This reduces energy consumption while taking advantage of the thousands of incandescent luminaires. (Of course, installing energy efficient lamps into existing incandescent fixtures does not qualify as an energy-efficient luminaire.) Choose sources that have the same lumen rating as the incandescent lamps for which the luminaire was designed.

Diffusers

Compact light sources—those typically used in small decorative luminaires—can create hot spots on glass and plastic diffusers. Deeper luminaires are less prone to such lamp images. Cased glass (multilayered) offers better diffusion than simple opal (plain white). Faceted, clear glass enclosures, designed for pleasing sparkle, should use clear filament lamps rather than coated ones. Choose lamps with low lumen output (and wattage) to minimize glare.

Maintenance

Lighting in kitchen and bath areas tends to operate longer than that in other rooms in the home. For this reason, it is worth considering the maintenance of decorative luminaires. In particular, how easy is it to replace spent light sources and electronic components, which typically are concealed by glass or plastic diffusers? Look for captive fasteners or diffusers that swing down for access to the interior.

Fluorescent Luminaires

Fluorescent luminaires offer the benefits of energy-efficient, diffused lighting, which is particularly useful in task areas.

Strip Lights

Strip lights—those used in coves and under shelves—are available in a variety of sizes, which may be a concern in compact spaces depending on the standard lengths in which the strips are available.

- *Length*. Depends on the size of the lamp. Using luminaires and lamps of the same length (e.g., rather than mixing 4 foot and 3 foot) not only simplifies maintenance but also improves the consistency of the lighting.
- *Width*. Depends on the number of lamps. Narrow versions, notably for use with T5 lamps, are also available.
- *Height*. Depends on the type of lamp (T5 strips are shorter than T8) and whether the lamp holders are on the top of the strip or the side.

Lamp Color

Generally, people prefer the warm tone of 2700 K or even the cooler 3000 K fluorescent lamps. And while 4000 K might feel appropriate in a utility space, introducing a contrasting tone in adjacent spaces can feel awkward and distracting. *Good CRI (80+) is a must* in kitchens and bathrooms and other areas of the home where color is an important factor and is readily available in either T8 or T5 lamps.

To ensure color consistency from luminaire to luminaire, fluorescent lamps should be from one manufacturer. Each manufacturer has its own phosphor "recipe"; all the recipes conform to American National Standards Institute standards, but they are not precisely the same. (Since compact and linear lamps use different phosphor mixtures, they can be from different manufacturers. Do not expect a 3000 K linear fluorescent luminaire to appear exactly the same as a 3000 K CFL luminaire. It is also a good idea to replace all of the lamps at one time. This way, you can ensure color consistency, and any lumen deterioration will be consistent across the entire area.

Dimming

You can dim fluorescent luminaires provided they include a dimming ballast (which you control with a fluorescent dimmer). Many manufacturers offer dimming options on their CFL downlights, although these carry an added cost and perhaps a longer delivery time that may not work for your project schedule.

Dimmable linear fluorescent strip lights and decorative luminaires may be more difficult to find. Since dimmable ballasts are generally the same size as nondimmable ones and often are available from distributor inventory, a good electrician may be able to rewire fixtures in the field for dimming.

LED Luminaires

As the latest technology on the lighting scene, LED luminaires pose several distinctive issues in regard to quality, color, dimming, and light output.

Quality

There is still much we have not pinned down about LED performance over time and in variable conditions. So product quality—broadly defined—has even more impact than it does with other luminaire types. Look for manufacturers with a track record in the technology and a reputation for standing behind their products.

Since you cannot anticipate every contingency, rely on partners who have demonstrated an ability to think problems through in advance and address the unanticipated when it occurs.

Color

LED color definition and consistency can vary more than fluorescent. It is best to use a single manufacturer for each type of luminaire to minimize color variation. And ask if similar products use identical LED components.

Do not expect the color appearance of LED luminaires to match those of incandescent and fluorescent luminaires, even at the same Kelvin values.

Dimming

You can dim LED luminaires provided they use a dimming driver (and you control them with a compatible dimmer). Many LED luminaires now feature dimmable drivers as standard equipment, eliminating the special-order issues typical of fluorescent luminaires. Just be certain.

Light Output

The light output (lumens) of a dedicated LED luminaire is measured directly, and the result is easily found on the product data sheets. The light output of incandescent and fluorescent luminaires, by contrast, depends on the specific light source used and is always less than the lumen rating of the lamps. This complicates the evaluation of luminaires with different sources. We discuss how to estimate lighting results in the next section, but you need be careful to make apples-to-apples comparisons.

Accent Luminaires

The term "accent luminaires" generally refers to those that can be adjusted or aimed to illuminate specific objects. Light source color, beam spread, and intensity are critical variables.

When accenting objects with light, choose a light source that will cover the object as intended from the appropriate location. Use manufacturer's charts to find the lamp that will provide proper beam spread, color, and intensity. Remember, you can arrange the beam to cover the entire object and contain it altogether or you can focus the beam on a particular part of the object. The following are two types of luminaires typically used for accent lighting.

Recessed Adjustable Luminaires

In selecting recessed adjustable luminaires, look for these characteristics:

- *Tilt angle.* Look for a minimum of 30 degrees (from vertical) and as much as 45 degrees for optimum aiming.
- *Shielding.* Unless the light source is well shielded by its location or a concealed pin-hole faceplate, accent luminaires tend to be quite glary.
- *Aperture size.* Accent luminaires arranged with simple downlights probably will look best if they use the same aperture size. When a single accent luminaire is used, the aperture can be any dimension that provides the appropriate tilt.

Track-Mounted Luminaires

Track-mounted luminaires provide the most flexibility for location and the most adjustability for aiming. The electrical track also solves many problems of electrical access.

Where track is used extensively, consider two-circuit versions that offer better switching and dimming control. Avoid using track, with its exposed conductors, in the damp areas of a bathroom. If using track on top of a beam, arrange the track to face sideways, not upside-down. This reduces the risk of material falling into the conductors.

While the track luminaire design can be solely an aesthetic decision, you also should consider how well the luminaire shields the light source from view. The ability to further modify or shield the beam with lenses or louvers may be a valuable feature. Many luminaires will accept only one size of lamp (e.g., MR16 or PAR38), so light source choice may drive luminaire choice.

SIZING THE LIGHT SOURCE

The approach to design development we have followed places luminaires based on the location of tasks and the architecture of the space (rather than on an illuminance calculation). Location, rather than the amount of light, determines the number of luminaires. After all, *where* the light falls affects our perception at least as much as the quantity of light that is actually there.

Nevertheless, we want to be sure that the quantity is sufficient for the tasks that need to be performed. We look at four typical conditions:

1. Ambient illumination from downlight luminaires
2. Ambient illumination from a cove
3. Task illumination from under-cabinet task luminaires
4. Lighting for grooming at a vanity

In older texts, wattage was often the simple proxy lighting quantity. In the guidelines that follow, we size lamps and luminaires in lumens (not watts). Doing this enables you to move easily among technologies of different efficacies (e.g., incandescent versus LED).

Ambient Illumination from Downlights

We are looking for the illumination over an area and begin with the definition of illuminance from Chapter 4, "Seeing the Work": 1 FC equals 1 lumen (lm) per square foot.

Basic Method

To deliver the desired quantity of illumination (FC), you decide how many lumens are required for each luminaire. That is, you size the light source using this five-step process:

1. Determine the recommended illuminance (FC).
2. Measure the area (square footage) to be illuminated.
3. Calculate the number of lumens that must arrive on that area to deliver the recommended FC.
4. Once you have your lighting plan completed and the quantity of luminaires required, estimate the number of lumens that must be provided in each luminaire in order to deliver the quantity needed.
5. Select a light source that provides the desired quantity of lumens.

Steps 1 to 3 are simple; step 4 needs some explanation. The following section explains how the process works.

Sizing an Ambient Light Source

You want to illuminate a 20- by 20-foot kitchen for cleaning and general tasks. (You will provide supplemental illumination on key work surfaces.) You have decided to use small-aperture, low-voltage MR16 halogen downlights spaced 5 feet on center, a total of 16 luminaires.

1. For the types of tasks in this area, you determine that an average of 10 FC will suffice. (See Chapter 4.)
2. You calculate that the area to be illuminated in this way is 400 square feet (20 × 20 = 400).
3. You calculate the lumens needed at the surface as 10 lm arriving per square foot (SF). 10 FC = 10 lm/SF and 10 lm/SF × 400 SF = 4000 lm.

4. Now that you have determined how many lumens are required, you need to choose the lamp and luminaire suitable to achieve this illumination.

    a. You are using 16 luminaires, so this equals 250 lumens per luminaire (4000 ÷ 16 = 250).

    b. You estimate about 75 percent of the light that originates in each luminaire ultimately will reach the target surface. Where did 75 percent come from?

    c. Thus, you want to find a source that provides 333 lumens. (250 needed at the surface ÷ 0.75 arriving = 333 needed in the source). The more light that ultimately arrives, the less you need in the source; the less light that arrives, the more you need. Try the formula with losses of 20 percent and 60 percent—that is, 80 percent and 40 percent arriving.

5. Studying your lamp catalog, you find that a basic 35MR16/FL36 is rated at 400 lumens, while a high-performance version is rated at 540 lumens. You also see a high-performance 20 W lamp rated at 320 lumens. Any of these would work. Since your client values sustainability, you select the 20-W option.

Surprised? The manufacturer's literature for the downlight indicates that it is rated for a maximum of a 50-W lamp. And many designers might use that lamp instead of fine-tuning the specification. If you have not already done so, please perform all of the calculations yourself. It really helps.

Losses in the Luminaire and Room

In the previous example, we sized the light source by assuming that only 75 percent of its lumens would reach the target surface and 25 percent would be lost.

- How much light is lost within the luminaire before light exits depends on the optics of the luminaire and type of light source.
- How much light is lost in the room before light reaches the target surface depends on the reflectances in the room (dark versus light finishes; smooth versus textured surfaces) and how much light hits those surfaces on the way to the target.

For technical lighting design, you can research and estimate each of these variables for specific luminaires and the interior architecture. With these data you can arrive at a reasonably precise expectation of the percentage of light lost (and, by subtraction, the percentage arriving). You also can model the results by entering the data into various computer programs.

For most residential design—where the precise lighting quantity is less important than other lighting results—the effort at precision rarely pays off. Instead, Table 14.4 provides a highly simplified approach.

Determine the type of luminaire and light source you are using. Then select the percentage of light arriving based on the size of the room and reflectance of the room surfaces.

Note that the percentages for dedicated LED luminaires are higher than for most conventional luminaires. That is because the lumen ratings of LED luminaires already account for losses inside the luminaire.

TABLE 14.4 Sizing the Light Source for Ambient Illumination

Luminaire	Light Source	% Light Arriving*	Note
Downlight	A lamp or CFL	50–35	Including LED lamps
Downlight	PAR, BR, MR16	80–65	Including LED lamps
Downlight	LED dedicated**	85–70	
Decorative	Incandescent or CFL	40–20	Including LED lamps
Decorative	Linear fluorescent	50–35	
Decorative	LED dedicated**	85–70	

*Includes both luminaire and room losses. The percentage is applied to the lumen rating of the light source except for dedicated LED luminaires, where it is applied to the lumen rating of the luminaire.

**Use the high end of the range (% light arriving) for larger rooms and more reflective surfaces; use the lower end for smaller rooms and less reflective surfaces.

TABLE 14.5 Sizing a Cove for Ambient Illumination

Luminaire	Light Source	Lumens per Foot*	Note
1-lamp	F32T8/830	300–120	Dimmable is 10% higher
2-lamp	F32T8/830	550–220	Dimmable is 10% higher
1-lamp	F28T5/830	350–140	
2-lamp	F28T5/830	650–250	
1-lamp	F54T5HO/830	625–250	Where 2-lamp strip won't fit

*Includes both luminaire and room losses.
Both T8 and T5-lamp strips generally are needed only where the length of the cove is small relative the area to be lighted or where a high level of illumination is needed.
Use the high end of the range of lumens per foot for larger rooms and more reflective surfaces; use the lower end for smaller rooms and less reflective surfaces.

Ambient Lighting from a Cove

We use a similar concept for this application as that for ambient overhead luminaires.

1. Determine the recommended illuminance (FC).
2. Measure the area (square footage) to be illuminated.
3. Calculate the number of lumens that must arrive on that area to deliver the recommended FC.
4. Measure the length of your cove.
5. Estimate the number of lumens that must be provided per foot of cove in order to deliver the quantity needed.
6. Using Table 14.5, determine the lamping (1 or 2 lights, T5 or T8 needed).

Task Lighting from Undercabinet Luminaires

Many manufacturers provide application guidelines for their products, illustrating expected illumination from a typical installation.

Sizing a Cove

You want to illuminate a 20 × 15-foot kitchen area for cleaning and general tasks. (You will provide supplemental illumination on key work surfaces). You would like to create a cove that runs on both 15-foot sides.

1. For the types of tasks in this area, you determine that an average of 10 FC will suffice. (See Chapter 4.)
2. You calculate that the area to be illuminated in this way is 400 square feet (20 × 20 = 400).
3. You calculate the lumens needed at the surface as 10 lm arriving per SF. 10 FC = 10 lm/SF, and 10 lm/SF × 400 SF = 4000 lm.
4. You have 30 feet of cove.
5. With 30 feet of cove, you will need to deliver 133 lumens per foot (4000 ÷ 30 = 133).
6. Using Table 14.5, you determine that a single T8 strip light will provide the target illumination. Since each strip measures 4 feet, you will use four per side, overlapping them by 3 inches to fit in the 15-foot length. This will also serve to minimize socket shadows.

If the luminaire is placed closer to the surface, the illuminance on the task surface directly below increases (but the light will cover a smaller area). If it is farther away, it decreases but covers a wider area.

If you cannot find application data for the product you want, you can apply these guidelines:

- *LED task luminaires.* 200 to 250 lm/ft in *luminaire* output
- *Fluorescent task luminaires.* 400 to 500 lm/ft in *lamp* output

Multiply the counter length by the lumen values to find appropriately size luminaires.

Lighting for Grooming at the Vanity

For lighting mounted above the mirror, beside the mirror, or a combination:

- Use a minimum of 36 inches of linear luminaires (the more the better), *or*
- Use a total of 2500 to 3500 lumens (from all lamps), depending on the light transmission of the luminaire diffusers.

A cove or pocket with two fluorescent lamps typically will deliver sufficient indirect illumination for grooming, provided the bathroom reflectances are high.

Reading Manufacturer Charts

Consulting manufacturer charts is recommended when planning lighting. These charts are particularly useful because the data are difficult to estimate without a computer program.

CONTROLS

As we discussed in Chapter 13, "Lighting Controls," you should create a controls strategy along with the concepts for the overall lighting design. To develop that strategy into a complete design, you will:

- Create zones to be controlled together.
- Determine the load in each zone.
- Select the specific control devices.
- Locate the controls.
- Program scenes where applicable for control systems.

Creating Zones

A *zone* is a group of lights controlled together. Each layer of lighting and each different light source needs its own zone. You can identify the zones graphically on your lighting layout and list them in a schedule.

Here is an example of a partial list of zones in the kitchen:

A. General downlights
B. Under-cabinet task lights
C. Peninsula pendants
D. Breakfast nook downlights
E. Breakfast nook pendant

While you might control the lighting with fewer zones—combining downlights and pendants, for example—you significantly limit the ability to adjust the atmosphere for different activities or modes.

Determining Loads

Once you have identified the zones, you can identify the load on each one. You describe the load by the type of light source and the total wattage.

Continuing our example:

A. General downlights: LED-ELV, 9 × 20 W = 180 W
B. Under cabinet task lights: LED-ELV, 20 feet @ 6 W per foot = 120 W
C. Peninsula pendants: Halogen, 12V ELV, 4 × 20 W = 80 W
D. Breakfast nook downlights: LED-ELV 4 × 20 W = 80 W
E. Breakfast nook pendant: Incandescent, 3 × 43 W = 129 W

Selecting Specific Control Devices

Selecting specific controls includes choosing the functionality, linkage, and appearance as well as coordinating the controls capacity with the load requirements. Review Chapter 13 for a discussion of the options.

System or Stand-Alone Controls

With five zones and the benefits of linking the kitchen lighting control to that of the adjacent dining room, we will use a system of "smart" dimmers that can be controlled wirelessly by push-button keypads. Using stand-alone controls would be awkward and inconvenient.

Luminaire/Control Compatibility

In selecting the specific dimmers, it is critical to ensure that the light source and auxiliary gear in the luminaire (in this case, LED drivers and low-voltage transformers) are compatible with the dimmers and other control devices. In our example, the luminaire specifications indicate that the drivers are dimmable with electronic low-voltage dimmers, and the transformers in the pendants are electronic.

It is also important to verify that the setup complies with any minimum load requirements in the dimmer and that ganging the controls together has not reduced the capacity of the dimmer below the actual load. If so, select a dimmer with a higher load capacity.

Our list includes keypads as well as dimmers, with the dimmer type and load specified. The keypad has six buttons, enough for all on, all off, and four separate lighting scenes. (The catalog number prefix "SDW" in the list below is made up.)

A. General downlights: SDWDim300-ELV
B. Under cabinet task lights: SDWDim300-ELV
C. Peninsula pendants: SDWDim300-ELV
D. Breakfast nook downlights: SDWDim300-ELV
E. Breakfast nook downlights: SDWDim600
F. Keypad: SDWKey6

Ganging Dimmers

The term "ganging" refers to installing several dimmers next to each other in one multiple-device wall box (as opposed to installing each dimmer in separate single-device boxes). Because of their close proximity, ganged dimmers can heat up more than dimmers by themselves. For this reason, dimmer manufacturers may specify that ganged dimmers be derated to account for the additional heat buildup.

TABLE 14.6 Matrix of Typical Scene Settings

Button	Scene	Down	Task	Peninsula	Nook Down	Nook Pendant
1	All/Clean	High	High	High	High	Med
2	Breakfast	High	High	Off	High	Med
3	Prep	High	High	High	Med	Off
4	Social	Med	Med	Med	Low	Off
5	Night	Off	Low	Off	Off	Off
6	All off	Off	Off	Off	Off	Off

Locating the Controls

Typically, entryways are the most convenient locations for controls. With several stand-alone controls grouped together, it can be difficult to identify which dimmer controls which luminaires.

With a system, however, the keypads can be located at the entry, and individual dimmers can be located so they relate to the luminaires they control. Alternatively, the individual dimmers can be located out of sight altogether and all control can be handled by the keypads.

A. General downlights: at kitchen entry
B. Under cabinet task lights: at edge of counter
C. Peninsula pendants: next to peninsula
D. Breakfast nook downlights: nook entry
E. Breakfast nook downlights: nook entry
F. Keypads: at kitchen and dining entries (2)

Programming

Since we are using a system with keypads, we need to determine how each button on the keypad will control the system. Since the actual commissioning may be performed by the electrical contractor, it is critical to establish the desired settings during this phase of design.

Table 14.6 provides a template for whoever actually commissions the system. The values are hypothetical, of course.

Lighting Concepts

This exercise asks you to analyze the light sources in your home or workplace. Record your results.

1. Create a list of fixtures that currently exist in your chosen space. How are they switched? Is it a successful placement of fixtures and design of switching? How would you improve the design?
2. Using Web or lamp catalog information, identify for each light source from question 1: the Kelvin temperature and color rendering index. Would you improve the lamping? How?

SUMMARY

Design development turns a lighting concept into a practical, buildable lighting design. Understanding the different light sources and how to use them allows you to put together the specifications required to make it more than an idea. Combining architectural details while keeping code compliance and architectural integration in mind ensures that your project is a successful one from the perspective of the individuals building it and the end users living in it. In Chapter 15, we discuss what documentation is required for implementation.

REVIEW QUESTIONS

1. What are the two codes that govern residential lighting? (See "Code Compliance" page 214)

2. What is the most sustainable choice for your lamping needs? (See "High Efficacy Luminaires" page 217)

3. When highlighting objects inside a cabinet, where is the best location for the lighting to be installed? (See "Lighting inside Cabinets" page 221)

4. What are four issues to consider with LED luminaires? (See "LED Luminaires" page 234)

5. Where are the three areas where lighting should be addressed in a bathroom? (See "Bath Lighting" pages 224–227)

Documenting the Lighting Design

Now that you have all your lighting concepts in mind, it is important to get them on paper. Having documentation that all consultants and contractors can follow is important for a successful project. A reflected ceiling plan with switching should be included as part of your drawing package. Details on how to implement your specific ideas as they relate to lighting are also illustrated on your reflected ceiling plan. This ensures that your design is executed correctly during the construction process. The location of lighting on the plan must be cross-referenced to schedules and specifications that give more detail on what you are asking for. This chapter gives you the information you need to be able to apply your ideas and provide necessary construction documents required for your project.

> *Learning Objective 1: Design a plan showing location and quantities of fixtures as well as the controls to operate them.*
>
> *Learning Objective 2: Communicate any specific lighting effects you want to achieve by crafting detailed sketches for implementation of your ideas.*
>
> *Learning Objective 3: Create a complete lighting specification package with information required for supply and installation of lighting and controls.*

LIGHTING AND CONTROLS PLANS

Once you have completed the programming and conceptual/schematic design part of the design process, you are ready to develop your ideas further. Getting your ideas on paper gives you the opportunity to see if concepts are viable and enables you to provide documentation for the next phase—construction. In a drawing package, a reflected ceiling plan usually follows the demolition/construction plan, which shows walls, windows, and doors. The reflected ceiling plan is drawn at the same scale and in the same orientation of the construction or floor plan. In addition, it follows similar line weight conventions.

A reflected ceiling plan (see Figure 15.1) can illustrate many components of the project (see Figure 15.2). These include:

- Shape of the ceiling, usually defined by walls
- Any bulkheads, coffers, or other details
- Type of ceiling material

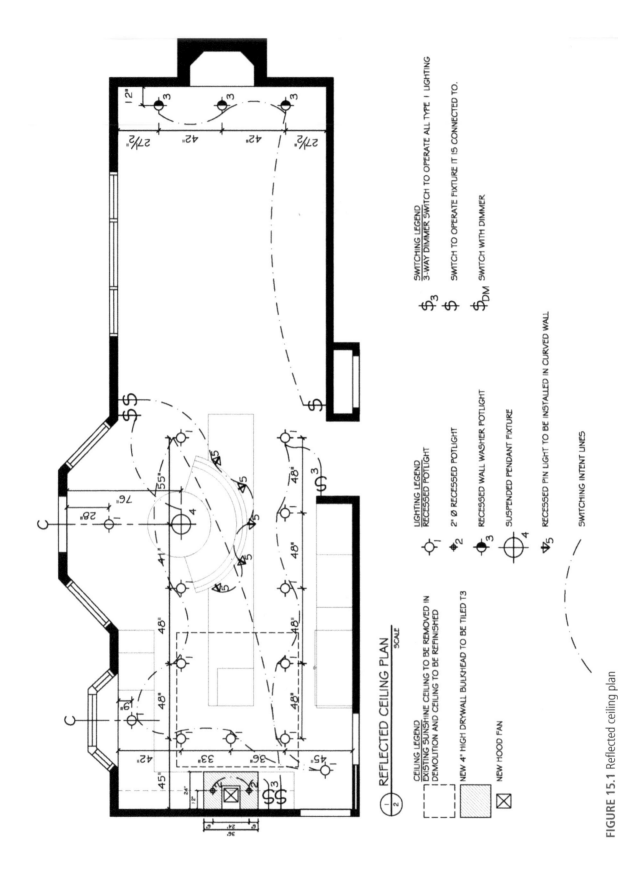

REFLECTED CEILING PLAN
SCALE

CEILING LEGEND
EXISTING SUNSHINE CEILING TO BE REMOVED IN DEMOLITION AND CEILING TO BE REFINISHED

NEW 4" HIGH DRYWALL BULKHEAD TO BE TILED T3

NEW HOOD FAN

LIGHTING LEGEND
RECESSED POTLIGHT

2' Ø RECESSED POTLIGHT

RECESSED WALL WASHER POTLIGHT

SUSPENDED PENDANT FIXTURE

RECESSED PIN LIGHT TO BE INSTALLED IN CURVED WALL

SWITCHING INTENT LINES

SWITCHING LEGEND
3-WAY DIMMER SWITCH TO OPERATE ALL TYPE 1 LIGHTING

SWITCH TO OPERATE FIXTURE IT IS CONNECTED TO.

SWITCH WITH DIMMER

FIGURE 15.1 Reflected ceiling plan
Design by Tammy MacKay, AKBD /Kim Van Ruskenveld, AKBD, Design Eye Ltd., Edmonton, AB

FIGURE 15.2 The lighting design in this kitchen was based on the reflected ceiling plan seen in Figure 15.1. *Design by Tammy MacKay, AKBD/Kim Van Ruskenveld, AKBD, Design Eye Ltd., Edmonton, AB*

- Height(s) of ceiling
- Location and type of light fixtures
- Switches and switching intent lines
- Lighting specifications
- Skylights and openings
- Heating, ventilation, and air conditioning components, if they exist

A separate plan, called electrical plan, is created showing outlets for electrical, data, and telephone requirements. This plan often shows furniture and how these items relate.

Details

In earlier chapters, we discussed different types of lighting effects. It is easy to describe what you want the end result to look and feel like, but how do you get there? A sketch or detailed drawing will help your idea become a reality. How do you get the concept of indirect lighting over top of a fireplace implemented unless you have a drawing to describe it? (See Figure 15.3.)

Figure 15.4 gives the general idea on how to build a bulkhead and cove in a space. It locates the light fixture as well. It is vital that the contractor is aware of these details early on in the project, so the electrical component of the lighting can be allowed for.

Dimensions

During the design process, you initially acquire all the information from your client regarding furniture and equipment. You have the final picture in mind of the space. The contractor that is installing the junction boxes and wiring is not familiar with what size and shape the dining room table is and whether a matching hutch will be added. It is up to you to ensure that the lighting is installed in the correct location. You must dimension the lighting from two reference points. Dimensioning to the center of the fixture will ensure that junction boxes are installed in the correct location. This also is important as symbols used to reference lighting are often not drawn to scale (see Figure 15.5).

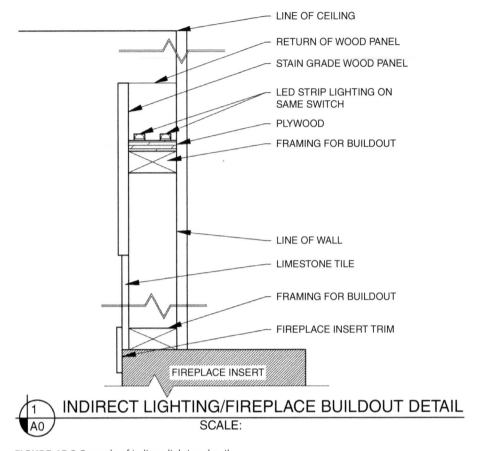

FIGURE 15.3 Example of indirect lighting detail
Design by Tammy MacKay, AKBD, Design Eye Ltd., Edmonton, AB

A lot of information is required to construct a space. Often plans can be overwhelmed with information, and small details cannot be seen at such a small scale. A larger-scaled drawing that is a portion of the complete reflected ceiling plan is provided with more detailed information. See the sketch in Figure 15.6 of a coffered ceiling. This sketch shows many details such as:

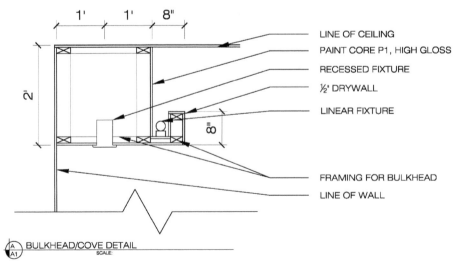

FIGURE 15.4 Example of cove detail
Design by Tammy MacKay, AKBD, Design Eye Ltd., Edmonton, AB

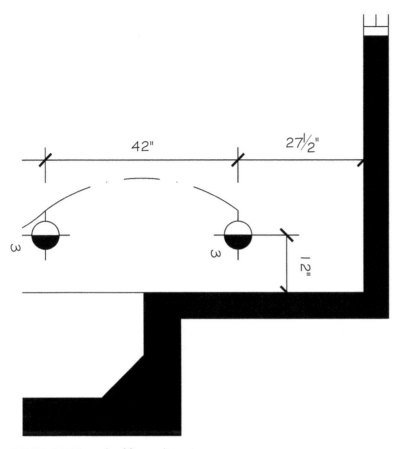

FIGURE 15.5 Example of fixture dimension

- Size of coffers
- Heights of coffers and surrounding borders
- Finish details of coffer
- Location of lighting in coffers

Specifications

Specifications are essential components of a drawing package. Specifications can be proprietary or generic. They provide all parties involved with specific information on products and materials used in the project. They give all the information required for ordering and installation. You can have both generic and proprietary specifications on a single project. Depending on your functional and aesthetic goals, either specification may be suitable. The following descriptions give information why you would choose one over the other.

Generic Specification

A generic specification allows you to give information on the performance, function, and general finish without actually choosing a specific one. For example, if you would like a 4-inch recessed fixture that has white trim and housing, you could say simply that. Electrical contractors have accounts with certain wholesalers that they prefer to work with, so if the brand is not important for functional or aesthetic reasons, you can give a generic specification of the fixture. Specific companies can be listed as prequalified manufacturers. This ensures a quality product and a competitive price.

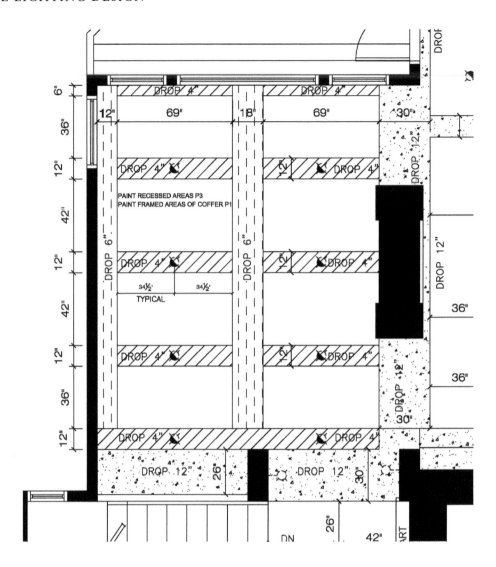

FIGURE 15.6 Coffer detail

Proprietary Specification

A proprietary specification is used when you have an exact vision in mind. You are saying that no substitution is permitted. In this case, you would show a specification that includes the manufacturer and model number of the specific fixture desired with specific finish information and codes.

Following are the components of a comprehensive specification:

Description of luminaire (fixture). Describes what kind of fixture it is. It could be a recessed fixture, a pendant fixture, or a wall-mounted fixture. It may be necessary to describe it as "NEW" because some projects may reuse some existing fixtures.

Manufacturer. When you have a specific fixture in mind, you would note who the manufacturer is.

Vendor. Give the contact information where the fixture can be purchased from. This helps the individual who is ordering the fixtures, as he or she may not be aware of the manufacturer or who represents the line of fixtures.

Model number. A model number is important for ordering purposes. Some manufacturers have model numbers that actually identify their product. It is a series of numbers and letters that ensures each product has a unique number. For example, the model number of the fixture seen in Figure 15.7 is based on this information:

FIGURE 15.7 Ventana fixture - 700WSVNTZLED
Courtesy of Tech Lighting

700WSVNT	FINISH	LAMP
	Z ANTIQUE BRONZE	IN INCANDESCENT 120V
	S SATIN NICKEL	CF COMPACT FLUORESCENT 120V
		LED 2700K 120V

Size. The model number should capture the sizing, but noting the size is important if any items located nearby may be impacted by the size of a surface-mount fixture. These include, for example, recessed fans or ducting located in the ceiling space.

Finish. A number of different finishes can be required for one fixture, such as frame finish, shade finish, and cord finish. Typically, a color or finish code is needed to ensure that the correct item is ordered. The finish codes typically are shown as letter codes derived from the name. Chrome is CH; oil-rubbed bronze is ORB; and satin nickel is SN. There are numerous finishes for lighting. Each manufacturer has finishes and terms that are unique to its products.

Location (also on plan). It is helpful to suggest where the fixture is being installed—for example, main floor powder room. There can be many fixtures in one residence, so cross-referencing the plan with the written specification is helpful.

Mounting height. There are no standard heights for installing suspended fixtures, only guidelines. There may be furniture under the fixtures that the electrician is not aware of. It is recommended to suggest hanging fixture at a specified distance above finished floor (AFF) to the bottom of the fixture.

Weight. Weight is often left off a specification. It becomes a valuable piece of information when the fixture is very heavy and requires more construction framing to brace its weight. For example, a foyer fixture may be oversized to fit into a two-story space. Weight also can be important for freight or shipping purposes.

Lamping. It is crucial to identify the number of lamps required, type of lamps, their wattage, and their Kelvin temperature. With the choices of artificial lighting and differences, these specific details can emphasize your finishes or take away from them. Differences in quality of lamps can affect pricing as well.

Schedules

A schedule organized in a table format explains the information you are passing on and makes it easy to follow. The schedule is used as a reference for ordering as well as installation

			LUMINAIRE SCHEDULE				
TYPE	SYMBOL	DESCRIPTION	MANUFACT./ MODEL #	LAMP	MOUNTING	QUANTITY	VENDOR
1		NEW LED WALL FIXTURE	EUREKA LIGHTING MODEL: SPIN LIGHT CODE: 3036-WH SIZE:10" Ø	22W LED KELVIN: 3000	TO BE MOUNTED AT 66" TO CENTER OF FIXTURE.	2	COMPANY ABC CONTACT: LES MULLIGAN @555-7000
2		NEW WALL FIXTURE	VIBIA LIGHTING MODEL: MILLENIUM CODE: 8090-01 FINISH: CHROME SIZE: 27" WIDE	24W T5 KELVIN: 3500	TO BE MOUNTED AT 66" TO CENTER OF FIXTURE.	3	COMPANY XYZ CONTACT: BETTY SMITH@ @555-1234
3		NEW FLUORESCENT WALL FIXTURE	VIBIA LIGHTING MODEL: MILLENIUM CODE: 8091-01 FINISH: CHROME SIZE: 38 3/4" WIDE	29W T5 KELVIN: 3500	TO BE MOUNTED AT 84" TO CENTER OF FIXTURE.	2	COMPANY XYZ CONTACT: BETTY SMITH@ @555-1234
4		NEW LED PENDANT	EUREKA LIGHTING MODEL: STACK CLUSTER CODE: 4436	1W LED KELVIN: 3000	TO BE MOUNTED AT 54" TO BOTTOM OF FIXTURE.	1	COMPANY ABC CONTACT: LES MULLIGAN @555-7000
5		NEW LED PENDANT	EUREKA LIGHTING MODEL: BEAN CODE: 4213-YFQ-CHR-ECR	8W LED KELVIN: 3000	TO BE MOUNTED AT 66" TO BOTTOM OF FIXTURE.	3	COMPANY ABC CONTACT: LES MULLIGAN @555-7000
6		NEW RECESSED LIGHTING	CONTRAST LIGHTING MODEL: A2000-01W WHITE TRIM & HOUSING	50W PAR 20 KELVIN: 3500	AS PER REFLECTED CEILING PLAN	12	COMPANY ABC CONTACT: LES MULLIGAN @555-7000
7		NEW LED ROPE LIGHTING	ARANI LIGHTING CODE: AR_LEDST5050-72W	KELVIN: 2700	AS PER REFLECTED CEILING PLAN	2	COMPANY ABC CONTACT: LES MULLIGAN @555-7000

FIGURE 15.8 Example of schedule

of fixtures. The example seen in Figure 15.8 suggests some standard table headings to use in your luminaire schedule. These headings can be modified to suit your specific project.

Symbols

Symbols are used to provide cross-references from a reflected ceiling plan to a luminaire schedule. The symbols shown in the sample schedule seen in Figure 15.9 would have a plan referencing those same symbols.

In addition to fixture symbols, switching symbols also are shown. These symbols show not only the switch location but also details, such as three-way switching or dimmers added to switching. Note that a three-way switch would be shown in two locations on your reflected

SWITCHING SCHEDULE			
SYMBOL	TYPE	STYLE/COLOR	NOTES
$E	EXISTING SWITCH		EXISTING SWITCH TO REMAIN
$	NEW SWITCH	LUTRON NEW ARCHITECTURAL COLOR: LIGHT ALMOND 'LA'	INSTALLATION NOTES:
$3	NEW THREE-WAY SWITCH	LUTRON NEW ARCHITECTURAL COLOR: LIGHT ALMOND 'LA'	INSTALLATION NOTES:
$DM	NEW DIMMER SWITCH	LUTRON NEW ARCHITECTURAL COLOR: LIGHT ALMOND 'LA'	INSTALLATION NOTES:
	SWITCHING INTENT LINE		AS NOTED

FIGURE 15.9 Example of switching symbols

ceiling plan. One fixture or set of fixtures would be controlled from two different points. The exact location of these two points would be illustrated for easy reference. The symbols show information that is important for the electrician to know early on in the construction process.

Notes

Including notes on the reflected ceiling plan is a good idea. The notes can include general information or specific instructions.

Examples of notes are:

- Ceiling height is 9'-0". (This is not captured on the plan but is a very important piece of information.)
- Color and style of toggles and cover plates to be Lutron Designer Matte Eggshell.
- Height of outlets and switches.
- Contractor shall undertake all work in accordance with codes and regulations by local authorities having jurisdiction.
- Contractor to ensure all fixtures and supporting hardware and wiring meets current local government standards.

Getting Started

Seven steps on how to create a reflected ceiling plan to include in your drawing package are listed next.

1. Start with the background floor plan. Show architectural elements and any other components that may interface with your lighting and ceiling details; for example, show doors as headers.
2. Decide what lighting you are using.
3. Create symbols to represent lighting.
4. Decide location of lighting and switching.
5. Dimension lighting and ceiling details.
6. Include any details and sketches.
7. Add a schedule with specifications and notes.

Reflected Ceiling Plan and Specifications

This exercise involves the light fixture seen in Figure 15.10. Use your journal to record your findings.

1. Put yourself in the shoes of the person who is going to be ordering the light fixture shown. What would you need to know about this fixture so that you will receive exactly what you want?
2. Your role now is to install the same fixture. What further information do you require to make sure you execute the final outcome correctly?
3. In a separate exercise, create a reflected ceiling plan of the existing lighting in one of the rooms in your home or office. Show the location of fixtures, switching, and switching intent lines. Is there anything you would change in your space? For example, add dimmers, split fixtures so they are on separate switches, or provide multiple switching (three-way).

FIGURE 15.10 Pendant fixture
Courtesy of Kichler

SUMMARY

Being able to communicate your ideas by providing drawings and specifications ensures that your project runs smoothly. Good communication, both written and oral, allows the contractors and subtrades to understand the design intent as well as order required fixtures and supporting hardware in a timely fashion. It will minimize errors and changes that can impact the project's schedule and budget.

REVIEW QUESTIONS

1. What would be used as your base to create a reflected ceiling plan? (See "Lighting and Controls Plan" pages 243–245)
2. Why is it important to include the weight of the fixture in the specification? (See "Proprietary Specification" pages 248–249)
3. What is the purpose of a three-way switch? (See "Symbols" pages 250–251)

Getting Lighting Built

Lighting is an important element of design, as we would not see anything without it. In the initial programming phase of the design process, you have decided, along with the stakeholders, the lighting concept that is desired. The initial lighting concept allows you to have an estimated budget in mind. You then go on to specify your lighting and locate each fixture in your construction documents. Several processes and tasks must be executed after this to ensure that your original concept and subsequent choices become a reality. Kitchens and bathrooms can contain some of the most exquisite materials and finishes. Lighting these in a way that shows their best characteristics ensures a space that is outstanding and loved by your client.

Learning Objective 1: Recognize when lighting is addressed during the project.

Learning Objective 2: Explain the sequence of events during the construction process as it pertains to lighting.

Learning Objective 3: Gain an appreciation of the many people involved in the lighting component of the project from the design phase through to project completion.

Learning Objective 4: List the required documentation from estimates to submittals.

LIGHTING IN THE CONSTRUCTION PROCESS

Lighting and components of lighting show up at all stages of the construction process. Lighting should be ordered so deadlines and schedules are met. Some lighting takes weeks or even months to arrive. Most lighting suppliers carry only about 10 percent of what they offer, so it is vital to ensure that lighting orders are placed in a timely fashion. You do not want to be forced into a situation where you must choose an alternate that does not meet your aesthetic or functional criteria.

Once construction has started and you are at the framing stage (see Figure 16.1), the electrician installs the wiring required to operate fixtures, switches, and outlets outlined in the construction documents. Junction boxes that surface fixtures will be attached to are also installed at this time. In addition, housing and required components are mounted for any recessed lighting that has been specified for the project. At this stage, all fixtures, locations, and details must be finalized.

FIGURE 16.1 Framing stage of construction

Once the wiring and installation of recessed components are complete, it is worthwhile to visit the site with your client and walk through the space for a final confirmation. You and your client know best where furniture and equipment will eventually be located.

You must pay special attention to these items:

- Switching
- Suspended fixtures
- Door swings

Switching

Ensure switching is not located behind future door swings or where intended tall pieces of furniture are to be located. Details such as these can easily be overlooked at this stage of construction, when only the studs are in place (see Figure 16.2).

Some switches may require a different and specific height depending on the location and function.

Suspended Fixtures

In regard to suspended fixtures, the likely approach from an electrician is to center a fixture in a room if only one is required. Furniture arrangements may, however, suggest that the center is not appropriate. For example, a dining room table may have a coordinating sideboard making it necessary to locate the dining table farther from the wall where the sideboard is located. In this case, putting the fixture in the middle of the ceiling would not give it the correct location over the dining table. The furniture placement and cabinet layout is information needed for correct placement of lighting.

Door Swings

The mounting height to the underside of the fixture cannot interfere with any doors and the operation of them. This refers to both doors for human entry and egress and for cabinet doors.

FIGURE 16.2 Installing switching at framing stage of construction

These special situations can vary depending on the project. It is important to have the end result in mind at each stage of construction.

As each stage is completed, making changes becomes more costly. Regular site meetings will ensure that any errors will be caught at an early stage. It may be necessary to discuss with the contractor and the electrician lighting details, such as switch and receptacle placement in regard to truss, beam, joist, and wall stud locations. Original plans may have to be modified if structural components have to be worked around.

One of the electrician's final tasks is the installation of the fixtures selected for the space. This is done after most of the subtrades have finished their parts of the project. At this stage, you

FIGURE 16.3 Completed kitchen
Design by Tammy MacKay, AKBD, Design Eye, Ltd., Edmonton, AB

will not have to worry about painters, cabinet installers, or other workers soiling or damaging fixtures that have taken weeks to arrive (see Figure 16.3).

SUBMITTALS

Submittals are documentation gathered in the project by various people and given to the architect or designer. They can be gathered by the general contractor or subcontractors (e.g., millworker or electrician) or can come directly from the vendor of the product itself.

They describe in detail the products, materials, and manufacturing or installation techniques used in the project. They are included as part of the construction documents. Submittals may include:

- Technical data sheets, which originate from the manufacturer (see Figure 16.4).
- Maintenance information, which originates from the manufacturer.
- Shop drawings, which originate from the subcontractors.
- Finish samples, which originate from the subcontractors. The actual sample given can vary considerably from the printed sample selection that was used for finish selection.

Once the submittals are reviewed, a sign-off is required. The sign-off is done by the architect or designer. The review of the submittals can take a few minutes or can be quite time consuming. It is an important step, especially if there are alternate products or manufacturing methods suggested by a subcontractor. The designer must review and approve with the original design concept in mind. The end product has to meet aesthetic criteria as well as functional goals. Once the designer approves the submittals, manufacturing of custom items continues and items are ordered. The designer now takes responsibility for the end results.

The general contractor typically manages the submittal process. He or she and the subcontractors must make this task a priority, as there can be lengthy lead times for product as well as manufacturing of custom pieces.

LIGHTING SUPPLY CHANNEL

A number of people may be involved in the lighting component of a project.

Clients

Clients are the most important stakeholders in the project. They have lighting requirements that must meet both functional and aesthetic goals. In the initial programming phase, clients must be honest about what they want to spend on lighting, so that the design concept presented meets with their suggested budget. Keeping clients informed and getting approval of submittals along the way is a necessary part of the process.

Designer

The designer is often the primary consultant who acts as a liaison between clients and all the other people involved in the project. When it comes to lighting, the designer also is responsible for:

- Collecting all the information from clients regarding needs and wants and researching the appropriate fixtures.
- Sourcing pricing within the budget and determining suitability and availability of initial lighting choices. Pricing can be received from agents or manufacturer representatives as well as retailers, depending on professional relationships, which can be unique to each design firm.
- Providing complete specifications and pertinent information required for the supply and installation of the lighting.
- Receiving, reviewing, and approving submittals provided.

Designers need to know what clients would like to spend on lighting during the programming phase of the project. This will affect all lighting decisions to come. Lighting typically is installed toward the end of the project. If the designer has done his or her job correctly, the importance of lighting to the overall design has been impressed on clients. In new construction and renovation, client patience and budget often have been taxed toward the end of the project. It is important that they still are committed to and satisfied with the end result that the designer painted a picture of in the beginning.

Consultants such as engineers (electrical, mechanical, and structural), home automation specialists, ergonomists, and occupational therapists also may be engaged in a project. Their involvement occurs during the design development prior to construction. They work as a team with clients and designer, so that all details requiring special attention are refined by an expert in the related field. In some cases, local authorities who have jurisdiction will require reports and approvals from such professionals.

General Contractors

Once all of the drawings and specifications are complete, you are ready to engage a general contractor. When you have a good set of project documents, it is easy to get cost estimates on the project, knowing you are comparing apples to apples. If specifications are incomplete or confusing, pricing can be equally incomplete and confusing. Once you award the project, a contract typically is drawn up between clients and the contractor. Consultants should not act as contractors; typically they do not carry appropriate insurances and other necessary accreditation related to construction.

The general contractor is the individual who hires the subtrades required for the project. He or she manages the construction and closely monitors schedule, costs, and workmanship.

Electricians work for the general contractor and are responsible for ordering the fixtures and controls. They can source any fixture but may have accounts with certain wholesalers where

Date: _____   Type: _____

Firm Name: _____

Project: _____

eW Profile MX Powercore
2700 K

Under-cabinet workplace LED task light for rapid retrofits and exceptional energy savings

eW Profile MX Powercore is a linear, direct line voltage LED under-cabinet light for common office and workplace task applications. Offers a rapid LED retrofit for traditional halogen or fluorescent task lights. Low power draw per foot, ENERGY STAR rating, and optional occupancy sensor / photosensor maximize energy efficiency.

- Versatile plug-and-play installation — A low-profile housing allows discreet under-cabinet placement. Fixtures are availble in 20 in (508 mm) and 40 in (1016 mm) lengths with surface mounting or magnet mounting options, and plug directly into US standard wall plugs.

- Dramatic energy savings — 20 in (508 mm) fixtures consume just 9 W, and 40 in (1016 mm) fixtures consume 18 W. Optional occupancy sensor for additional energy savings.

- A range of options for design and application flexibility — Available in 2700 K, 3000 K, 3500 K, and 4000 K color temperatures. White housing matches a range of office decors.

- Integrates patented Powercore technology — Powercore rapidly, efficiently, and accurately controls power directly from line voltage, eliminating external power supplies.

- Superior color consistency and accuracy — Optibin, an advanced binning algorithm, exceeds the recognized standards for color quality to guarantee uniformity and consistency of hue and color temperature.

For detailed product information, visit www.philipscolorkinetics.com/ls/essentialwhite/ewprofilemxpowercore/

Specifications   Due to continuous improvements and innovations, specifications may change without notice.

Item	Specification	20 in (508 mm)	40 in (1016 mm)
Output	Beam Angle	110°	110°
	Lumens†	337	714
	Efficacy (lm / W)	41.5	43.0
	CRI	84	85
	Lumen Maintenance‡	50,000 hours L70 @ 50° C	
Electrical	Input Voltage	120 VAC, 50 / 60 Hz	
	Power Consumption	9 W max at full output, steady state	18 W max at full output steady state
	Power Factor	.99 @ 120 VAC	.99 @ 120 VAC
Control		Integrated ON / OFF switch, optional occupancy sensor	
Physical	Dimensions (Height x Length x Depth)	1.7 x 18.4 x 2.5 in (43 x 468 x 63.5 mm)	1.7 x 36 x 2.5 in (43 x 914 x 63.5 mm)
	Weight	2 lb (.91 kg)	3.1 lb (1.4 kg)
	Housing	White painted metal	
	Lens	Diffuse polycarbonate	
	Fixture Connections	Standard wall plug	
	Temperature Ranges	-4° – 122° F (-20° – 50° C) Operating -4° – 122° F (-20° – 50° C) Startup -40° – 176° F (-40° – 80° C) Storage	
	Humidity	0 – 95%, non-condensing	
Certification and Safety	Certification	UL / cUL, FCC Class A, ENERGY STAR	
	Environment	Dry / Damp Location, IP20	

† Lumen measurement complies with IES LM-79-08 testing procedures.

‡ L70 = 70% lumen maintenance (when light output drops below 70% of initial output). Ambient luminaire temperature specified. Lumen maintenance calculations are based on lifetime prediction graphs supplied by LED source manufacturers. Calculations for white-light LED fixtures are based on measurements that comply with IES LM-80-08 testing procedures. Refer to www.philipscolorkinetics.com/support/appnotes/lm-80-08.pdf for more information.

Fixtures

Item	Color Temp.	Length	Item Number
eW Profile MX Powercore Surface Mount	2700 K	20 in (508 mm)	523-000068-00
		40 in (1016 mm)	523-000068-08
	3000 K	20 in (508 mm)	523-000068-02
		40 in (1016 mm)	523-000068-10
	3500 K	20 in (508 mm)	523-000068-04
		40 in (1016 mm)	523-000068-12
	4000 K	20 in (508 mm)	523-000068-06
		40 in (1016 mm)	523-000068-14
eW Profile MX Powercore Surface Mount with Occupancy Sensor	2700 K	20 in (508 mm)	523-000068-01
		40 in (1016 mm)	523-000068-09
	3000 K	20 in (508 mm)	523-000068-03
		40 in (1016 mm)	523-000068-11
	3500 K	20 in (508 mm)	523-000068-05
		40 in (1016 mm)	523-000068-13
	4000 K	20 in (508 mm)	523-000068-07
		40 in (1016 mm)	523-000068-15
eW Profile MX Powercore Magnet Mount	2700 K	20 in (508 mm)	523-000068-16
		40 in (1016 mm)	523-000068-24
	3000 K	20 in (508 mm)	523-000068-18
		40 in (1016 mm)	523-000068-26
	3500 K	20 in (508 mm)	523-000068-52
		40 in (1016 mm)	523-000068-28
	4000 K	20 in (508 mm)	523-000068-22
		40 in (1016 mm)	523-000068-30
eW Profile MX Powercore Magnet Mount with Occupancy Sensor	2700 K	20 in (508 mm)	523-000068-17
		40 in (1016 mm)	523-000068-25
	3000 K	20 in (508 mm)	523-000068-19
		40 in (1016 mm)	523-000068-27
	3500 K	20 in (508 mm)	523-000068-21
		40 in (1016 mm)	523-000068-29
	4000 K	20 in (508 mm)	523-000068-23
		40 in (1016 mm)	523-000068-31

Photometrics

eW Profile MX Powercore
2700 K, 20 in (508 mm)

Polar Candela Distribution

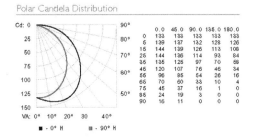

Illuminance at Distance

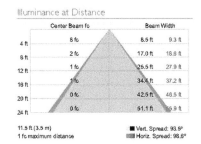

For lux multiply fc by 10.7

Lumens	337
Efficacy	41.5 lm / W

2700 K, 40 in (1016 mm)
Polar Candela Distribution

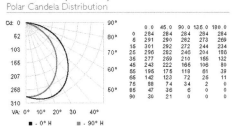

Illuminance at Distance

For lux multiply fc by 10.7

Lumens	714
Efficacy	43.0 lm / W

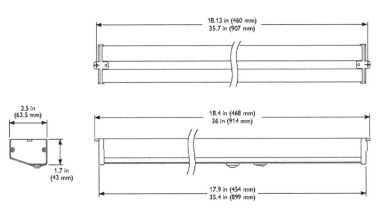

Philips Color Kinetics
3 Burlington Woods Drive
Burlington, Massachusetts 01803 USA
Tel 888.385.5742
Tel 617.423.9999
Fax 617.423.9998
www.philipscolorkinetics.com

FIGURE 16.4 Example of submittal by Philips

they may get preferred pricing. As a result, they may suggest alternate fixtures. In such cases, they present the designer with a submittal to review. If the original design intent is met, the designer may approve the alternate fixtures suggested.

Manufacturers

Manufacturers are the companies that make the lighting and any supporting hardware and components. Manufacturers are located all over the world. It is vital that lighting is ordered in a timely fashion, since it can take up to 16 weeks for your order to arrive, depending on where it is coming from.

Manufacturer's Representative

A manufacturer's representative is an individual employed by the manufacturer to promote its products to wholesalers. In some cases, the representative works directly with specifiers, such as interior designers or lighting designers. A manufacturer's representative also can work directly with large home improvement retailers. With no middleman involved, pricing can be less. Manufacturer's representative may have limited offerings from the manufacturers they represent, so not all pieces in a catalog may be available for purchase through this type of retailer.

Agents

Agents are very similar to manufacturer's representative (see Figure 16.5). Unlike a manufacturer's representative, they are not employed by the manufacturer and typically are compensated on a commission-based structure. This allows them to be agents for multiple manufacturers. An agent promotes the manufacturers' products to wholesalers.

Both agents and manufacturer's reps also promote their products to architects and designers. These professionals are not sellers of the product, but they do require detailed information in order to create specifications for the projects on which they are consulting.

Wholesalers

Wholesalers carry products from numerous manufacturers. They do not have a showroom or carry samples for display. Wholesalers may, however, decide to stock certain items offered by a manufacturer. There may be a few items that they always keep in stock to meet demand, such as standard recessed downlights. For some projects with tight timelines, working with a wholesaler with certain stocked items may be your only option. Retailers purchase product from wholesalers. Electricians and some contractors also may purchase through wholesalers, provided they have accounts. They will then mark up the product and include labor in the pricing they submit to the general contractor or to the client (if they are working directly for the client). If there are any warranty issues, the wholesaler typically deals directly with the manufacturer to rectify the problem.

Retailer/Vendor

Retailers range from small boutique lighting shops, such as seen Figure 16.6, to large home improvement box-type stores. They sell directly to consumers. These businesses have show-rooms displaying a lot of products, although it is only a fraction of what they have access to. They also have catalogs available if what you want is not on display. You can order your lighting package directly from the retailer. It is helpful for clients to see what the end product will look like as well as the size and scale of the piece. It is also helpful for designers to see lighting finishes, especially if they are using multiple manufacturers. The showrooms

FIGURE 16.5 Lighting agent and designer

have the different finishes available for comparison. At a lighting showroom, salespeople may be able to help with lighting design. Some sales associates may have received formal training. You should confirm the sales associate's qualifications prior to your visit. You may have to make an appointment to get one-on-one attention. Electricians also can purchase lighting through retailers. They generally receive contract pricing and do not have to pay the suggested retail price that the typical consumer, walking in off the street, would be charged.

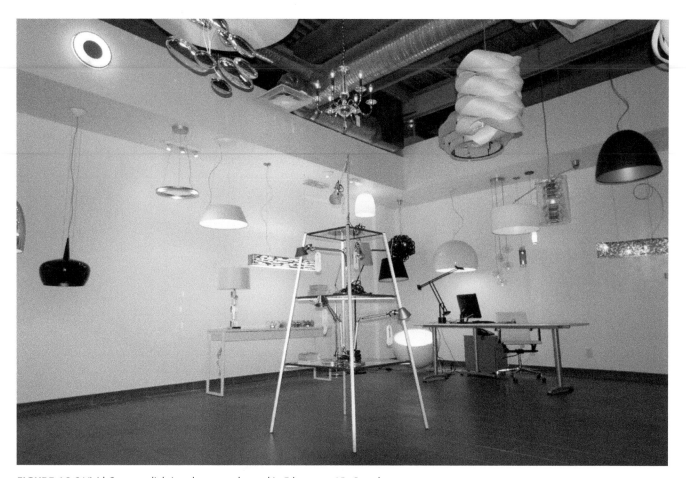

FIGURE 16.6 Vivid Concepts lighting showroom located in Edmonton, AB, Canada

When choosing a fixture, understanding the ordering process is important. You must know details such as who will be ordering the fixture and from where.

1. Research a fixture that you may want to use on a project. Find the technical data for this fixture. What specific decisions need to be made before the fixture can be ordered?
2. Take a look at your local lighting community. Identify who is an agent, manufacturer's representative, manufacturer, wholesaler, and retailer. List them in the order of the lighting supply channel described earlier.

SUMMARY

Construction administration during the project is necessary for a successful project. You must know the roles and responsibilities of all of the stakeholders. Keeping the flow of information constant and consistent will keep the lines of communication open and ensure that errors and forgotten details are kept to a minimum. At the end of the project, you will have a file complete with documentation that reflects decisions that were made along the way.

REVIEW QUESTIONS

1. Who typically manages the submittal process? (See "Submittals" page 256)
2. Who approves the submittals? (See "Submittals" page 256)
3. Give examples of submittals and who they would originate from. (See "Submittals" page 256)
4. Who purchases product from wholesalers? (See "Wholesalers" page 260)

Glossary

A

Accent lighting
Directional lighting used to highlight and draw attention to something, such as artwork, architecture, or landscape.

AFF (above finished floor)
A dimension given for mounting suspended light fixtures.

Ambient lighting
Also referred to as general lighting. Lighting that provides an overall illumination that allows people to move about safely.

Ampere (amp)
Unit of electrical current. Rate of electricity flow and circuit capacity are stated in amperes.

ANSI Standards
The American National Standards Institute (ANSI) is an organization that oversees the development of consensus standards for products and services in the United States.

Aperture
The open area of a fixture that allows light to exit.

B

Ballast
A device that regulates the amount of electrical current going through a fluorescent lamp. The lamp requires more voltage to start initially. The ballast ensures that once the lamp is illuminated, the current is regulated so it does not overheat.

C

Canadian Electric Code (CEC)
A book of referenced standards developed to help protect human safety at home and at work.

Canadian Standards Association (CSA)
A respected authority on safety and performance standards and certification in Canada.

Candela
Unit of candle power.

CFL
Abbreviation for compact fluorescent lamp.

Color rendering index (CRI)
The most widely used metric for color rendering. Color rendering represents the light's ability to interpret color.

Color temperature
A measurement used to describe the color characteristics of light, usually described as being warm or cool. Color temperature is measured in Kelvins (K).

Cool colors
In color theory, green, blue, and violet are referred to as cool colors. They are often associated with ice and cool water.

D

Diffuse
Spread out, not concentrated in one area. When a surface is rough and textured, light will reflect in a more diffused and scattered way.

Driver
An item that controls a device. Like a ballast or transformer, a driver is used with the light-emitting diode (LED) source. It regulates the power supply to the fixture so that it has enough power to illuminate but not so much to cause it overheat or burn out.

F

Footcandle (fc)
Unit of illuminance. One lumen falling on 1 square foot equals 1 footcandle.

G

Glare
When bright light is directed or reflected, causing discomfort of the eye.

Grazing
When lighting is placed in such a way that the whole surface is lit rather than having a direct light that creates a high light on one area.

I

Illuminance
Light falling on a surface or object. Measured in footcandles (fc) or lux.

Illumination
The act of lighting something.

Incandescent
An artificial light source that uses a glowing heated wire filament to produce light. Traditionally warm light has a Kelvin temperature of about 2700 to 3000.

Indirect lighting
Results when light bounces off a surface before illuminating a space (i.e., cove lighting).

Kelvin (K)
Scientific unit of temperature. Color temperature is measured on the Kelvin scale, for example 3000 K.

Kilowatt-hour
One thousand watt-hours. A measure of electrical energy consumed.

L

Lamp
Proper term given to an artificial light source (*not* bulb).

Life cycle assessment
Defined by the US Green Building Council as an analysis of the environmental aspects and potential impacts associated with a product, process, or service.

Life cycle cost
Defined by the US Green Building Council as a process that looks at both purchase and operating costs as well as relative savings over the life of a product.

Line voltage
Voltage used directly from source. Usually 120 volts (V).

Low voltage
Requires the use of a transformer to step the voltage down to 12 volts.

lm/W or LPW
Lumens per watt.

LRV (Light reflectance value)
The amount of light reflected off a surface. LRV is expressed as a percentage.

Lumen (lm)
A measurement of the total amount of visible light emitted by a source.

Luminaire
An electrical piece of equipment used to produce artificial light by a particular lamp source.

Luminance
Brightness of the surface seen from our point of view.

Luminous flux
A measurement, in lumens, of the amount of light energy radiating from a source of light.

Lux
Metric unit of illuminance. One lumen falling on 1 square meter equals 1 lux. (One lux equals 10.76 footcandles.)

N

National Electric Code (NEC)
A set of standards that describe the safe installation of electrical wiring and related components.

P

Photometry
The measurement of visible light.

Plenum
An air space above the ceiling that typically accommodates heating and cooling ducting, cabling, or recessed lighting fixtures.

R

Reflection
Light bouncing off a surface.

Refraction
The bending of the light wave when it passes through two different mediums where the light either speeds up or slows down.

S

Solar heat gain
When sunlight passes through a window, the space's temperature will rise. This can be a positive effect in the winter months but an unwanted effect in hotter months or in warm climates.

Spectral power distribution (SPD)
A measurement that describes the power per unit area per unit wavelength of an illumination.

Spectrum
In lighting, refers to the colors that make up white light: red, orange, yellow, green, blue, indigo, and violet.

Specular
When a surface is smooth and shiny, the surface acts like a mirror and reflects light creating a highlight on the surface.

Sustainability
Involving methods that meet the needs of the present without destroying natural resources and compromising the ability of future generations to meet their own needs.

T

Task lighting
Provides the illumination required for a particular task, such as reading or preparing meals.

Three-way switch
A switch that operates a fixture from two different locations: for example, a switch at top of stairs and another at the bottom of stairs that operate the same fixture in the stairwell.

Topology
A set theory based on researched concepts.

Transformer
Similar to a ballast for fluorescent lamps, a transformer regulates the incoming electrical current that is being sent to a low-voltage lamp, such as halogen. It steps the current down to a lower voltage to ensure the lamp can initially start up but not overheat once it has started.

U

UL (Underwriters Laboratories)
Independent, not-for-profit product safety testing and certification organization.

V

Volt
Unit of electrical force. Residential voltage in the United States and in Canada is 120 volts.

VLT (visible light transmission)
The amount of visible light that is transmitted through glazing (glass). Also referred to as VT.

W

Warm colors
In color theory, red, orange, and yellow are referred to as warm colors. They often are associated with fire.

Wattage
Unit of electrical power. The energy a lamp consumes is measured in watts (W). A 50-W halogen suggests it uses 50 watts of energy per unit of time.

X

Xenon
A type of light source that falls under the incandescent umbrella. It has a good color rendering index and Kelvin temperature and is used mostly in under-cabinet or recessed millwork lighting.

Index